Lecture Notes in
Operations Research and
Mathematical Economics

Edited by M. Beckmann, Providence and H. P. Künzi, Zürich

12

Mathematical Systems Theory and Economics II

Proceeding of an International Summer School
held in Varenna, Italy, June 1 - 12, 1967

Edited by
H. W. Kuhn
Princeton University, Princeton, N. J./ USA

G. P. Szegö
University of Milano, Milano/Italy

1969

Springer-Verlag
Berlin · Heidelberg · New York

Library of Congress Catalog Card Number 70-81409 Printed in Germany. Title No. 3761

Contents

Part III Special Mathematical Problems

Part IV Special Applications

Lecture Notes in
Operations Research and
Mathematical Economics

Edited by M. Beckmann, Providence and H. P. Künzi, Zürich

12

H. W. Kuhn · G. P. Szegö, ed.

Mathematical Systems Theory and Economics II

Springer-Verlag
Berlin · Heidelberg · New York

The Lecture Notes are intended to report quickly and informally, but on a high level, new developments in mathematical economics and operations research. In addition reports and descriptions of interesting methods for practical application are particularly desirable. The following items are to be published:

1. Preliminary drafts of original papers and monographs

2. Special lectures on a new field, or a classical field from a new point of view

3. Seminar reports

4. Reports from meetings

Out of print manuscripts satisfying the above characterization may also be considered, if they continue to be in demand.

The timeliness of a manuscript is more important than its form, which may be unfinished and preliminary. In certain instances, therefore, proofs may only be outlined, or results may be presented which have been or will also be published elsewhere.

The publication of the *"Lecture Notes"* Series is intended as a service, in that a commercial publisher, Springer-Verlag, makes house publications of mathematical institutes available to mathematicians on an international scale. By advertising them in scientific journals, listing them in catalogs, further by copyrighting and by sending out review copies, an adequate documentation in scientific libraries is made possible.

Manuscripts

Since manuscripts will be reproduced photomechanically, they must be written in clean typewriting. Handwritten formulae are to be filled in with indelible black or red ink. Any corrections should be typed on a separate sheet in the same size and spacing as the manuscript. All corresponding numerals in the text and on the correction sheet should be marked in pencil. Springer-Verlag will then take care of inserting the corrections in their proper places. Should a manuscript or parts thereof have to be retyped, an appropriate indemnification will be paid to the author upon publication of his volume. The authors receive 25 free copies.

Manuscripts written in English, German, or French will be received by Prof. Dr. M. Beckmann, Department of Economics, Brown University, Providence, Rhode Island 0 29 12/USA, or Prof. Dr. H. P. Künzi, Institut für Operations Research und elektronische Datenverarbeitung der Universität Zürich, Sumatrastraße 30, 8006 Zürich.

Die Lecture Notes sollen rasch und informell, aber auf hohem Niveau, über neue Entwicklungen der mathematischen Ökonometrie und Unternehmensforschung berichten, wobei insbesondere auch Berichte und Darstellungen der für die praktische Anwendung interessanten Methoden erwünscht sind. Zur Veröffentlichung kommen:

1. Vorläufige Fassungen von Originalarbeiten und Monographien.

2. Spezielle Vorlesungen über ein neues Gebiet oder ein klassisches Gebiet in neuer Betrachtungsweise.

3. Seminarausarbeitungen.

4. Vorträge von Tagungen.

Ferner kommen auch ältere vergriffene spezielle Vorlesungen, Seminare und Berichte in Frage, wenn nach ihnen eine anhaltende Nachfrage besteht.

Die Beiträge dürfen im Interesse einer größeren Aktualität durchaus den Charakter des Unfertigen und Vorläufigen haben. Sie brauchen Beweise unter Umständen nur zu skizzieren und dürfen auch Ergebnisse enthalten, die in ähnlicher Form schon erschienen sind oder später erscheinen sollen.

Die Herausgabe der *„Lecture Notes"* Serie durch den Springer-Verlag stellt eine Dienstleistung an die mathematischen Institute dar, indem der Springer-Verlag für ausreichende Lagerhaltung sorgt und einen großen internationalen Kreis von Interessenten erfassen kann. Durch Anzeigen in Fachzeitschriften, Aufnahme in Kataloge und durch Anmeldung zum Copyright sowie durch die Versendung von Besprechungsexemplaren wird eine lückenlose Dokumentation in den wissenschaftlichen Bibliotheken ermöglicht.

Part III Special Mathematical Problems

Control vector fields on manifolds and attainability*)

by Felix ALBRECHT

This paper contains a coordinate-free proof of the following statement: if $(\xi, \underline{U})$ is a given control vector field on a finite-dimensional manifold of class C^3, then any control $u \in \underline{U}$ steering a point of the manifold to the boundary of its set of attainability satisfies the Pontryagin Maximum Principle. Similar results for control processes in Euclidean spaces (under various restrictions) have been proved by different methods (see e.g. [2], [3], [4]).

The method used here was developed in the course of long and very useful discussions with Robert ELLIS, whose many suggestions have been essential.

<u>1.</u> Let X be a finite-dimensional Hausdorff C^p-manifold, $p \geq 3$, and Ω a Hausdorff topological space. A continuous mapping $\xi: X \times \Omega \to T(X)$ is called a <u>controllable</u> family of vector fields on X with control space Ω, if:

1) the partial mapping of ξ with respect to the first variable is of class C^{p-1} on $X \times \Omega$,

2) for any $(x,\omega) \in X \times \Omega$,

$$\pi\xi(x, \omega) = x ,$$

where π is the canonical projection of $T(X)$ onto X.

Let $\xi : X \times \Omega \to T(X)$ be a controllable family of vector fields on X with control space Ω. We denote by $\underline{U}_r$ the class of all mappings $u: I_u \to \Omega$, where I_u is a compact interval of $\mathbb{R}$ depending on u, such that the limits to the right and to the left of u exist at each point of I_u. In this case the mapping

$$\xi_u : X \times I_u \longrightarrow T(X) ,$$

defined by $\xi_u(x, t) = \xi(x, u(t))$ is a (time-dependent) vector field on X, satisfying the usual existence and uniqueness conditions for its integral curves.

---

*) The author was partially supported by NSF Grants GP-6247 and GP-6325 and by Grant AF-AFOSR-359-66.

A subclass $\underline{U}$ of $\underline{U}_r$ is called <u>admissible</u> (for Ω), if it satisfies the following conditions:

C.1. If $u : I_u \to \Omega$ is in $\underline{U}$, $\omega \in \Omega$ and J is an interval of $\mathbb{R}$, then the mapping $v : I_u \to \Omega$ defined by

$$v(t) = \begin{cases} u(t) & \text{for } t \in I_u - J \\ \omega & \text{for } t \quad I_u \cap J \end{cases}$$

is also in $\underline{U}$.

C.2. If $u : I_u \to \Omega$ is in $\underline{U}$, its restriction to any compact subinterval of I_u belongs to $\underline{U}$.

C.3. If $u : I_u \to \Omega$ is in $\underline{U}$ and $s \in \mathbb{R}$, the mapping $w_s : I_u + s \to \Omega$ defined by $w_s(t) = u(t-s)$ is also in $\underline{U}$.

C.4. If $u : I_u \to \Omega$ is a mapping defined on a compact interval $I_u = [a, b]$ and if there exists a $c \in (a,b)$ such that the restrictions of u to $[a, c]$ and to $[c, b]$ are in $\underline{U}$, then u itself is in $\underline{U}$.

A <u>control vector field on</u> X (with control space Ω) is a pair $(\xi, \underline{U})$, where $\xi : X \times \Omega \to T(X)$ is a controllable family of vector fields on X and $\underline{U}$ is an admissible class for the space Ω. The elements of $\underline{U}$ are called the <u>controls</u> of $(\xi, \underline{U})$. A vector field ξ_u with $u \in U$ is called <u>admissible for</u> $(\xi, \underline{U})$.

Let $(\xi, \underline{U})$ be a control vector field on X with control space Ω. A point $x_1 \in X$ is called <u>attainable from a point</u> $x_0 \in X$, if there exists a control $u : [a, b] \to \Omega$ in $\underline{U}$ such that the integral curve α_u of the vector field ξ_u with initial condition (x_0, a) is defined on $[a,b]$ and $\alpha_u(b) = x_1$. In this case we say also that u <u>steers</u> x_0 <u>to</u> x_1.

The set $A(x_0)$ consisting of x_0 and of the points in X which are attainable from x_0 is called the <u>set of attainability</u> of the point x_0.[1)]

---

1) By adjoining an element to the space Ω and modifying accordingly the class $\underline{U}$ we can always assume that the zero vector field on X is admissible for $(\xi, \underline{U})$.

2. We assume that $(\xi, \underline{U})$ is a given control vector field on the manifold X, with control space Ω. It is easy to see that we may consider, without loss of generality, only controls $u \in \underline{U}$ of the form $u : [0, t_u] \to \Omega$. Let $x_0, x_1 \in X$ such that $x_1 \in A(x_0)$ and u be a fixed control steering x_0 to x_1. Then $\alpha : [0, t_u] \to X$ denotes the integral curve of ξ_u with initial condition $(x_0, 0)$ and $\alpha_*(r, s) : T_{\alpha(r)}(X) \to T_{\alpha(s)}(X)$ the isomorphism between tangent spaces induced by the flow of ξ_u for $r, s \in [0, t_u]$.

Consider the set S of points of continuity of the mapping $t \mapsto \xi(\alpha(t), u(t))$. By the uniqueness property of α we may suppose that

$$\xi(\alpha(t), u(t)) = \lim_{\substack{\tau<t \\ \tau\neq t}} \xi(\alpha(\tau), u(\tau))$$

for all $t \in (0, t_u)$ and that u is continuous at the points 0 and t_u.

DEFINITION. The convex cone K_0 generated in $T_{x_0}(X)$ by the elements $\alpha_*(s, 0) . \xi(\alpha(s), \omega)$, $-\alpha_*(s, 0) . \xi(\alpha(s), u(s))$ for $s \in S$, $\omega \in \Omega$ is called <u>the cone of attainability of</u> u <u>at</u> x_0. If $K_0 \neq T_{x_0}(X)$ the control u is called <u>extremal</u>.

It is easy to see that u is extremal if and only if there exists a non-zero linear form ψ on $T_{x_0}(X)$ such that

$$\psi(\alpha_*(t, 0) . \xi(\alpha(t),\omega)) = \sup_{\omega \in \Omega} \psi(\alpha_*(t, 0) . \xi(\alpha(t), u(t))) = 0$$

for all $0 \leq t \leq t_u$ (this is a well-known fact if $X = \mathbb{R}^n$).

THEOREM. <u>If</u> x_1 <u>lies on the boundary of the set</u> $A(x_0)$ <u>then the control</u> u <u>is extremal</u>.

The proof of this theorem requires certain preliminary considerations; some of the necessary facts will be stated below without proofs, which will appear elsewhere [1].

Let $\mu : \mathfrak{M}(\xi_u) \to X$ be the flow of the vector field ξ_u, where $\mathfrak{M}(\xi_u)$ is an open subset of $X \times I_u \times I_u$; note that the mapping $t \mapsto \mu(x, \tau, t)$ is the integral curve of ξ_u with initial condition (x, τ) and that for fixed τ, t the mapping $x \mapsto \mu(x, \tau, t) = \mu_{\tau,t}(x)$ is a diffeomorphism on an open set of X.

Let U_0 be a coordinate neighborhood of x_0 such that $\mu_{0,t}$ is defined on U_0 for each $t \in [0, t_u]$. Define two families of vector fields on U_0 by setting:

$$\xi_{\omega,s,\lambda}(x) = \lambda(d_x\mu_{0,s})^{-1} \cdot \xi(\mu_{0,s}(x), \omega)$$

for $x \in U_0$, $\omega \in \Omega$, $s \in S$, $\lambda \geq 0$ and

$$\eta_{s,\lambda}(x, \tau) = -\lambda(d_x\mu_{0,s})^{-1} \cdot \xi(\mu_{0,s}(x), u(-\lambda\tau + (1+\lambda s))$$

for $(x, \tau) \in U_0 \times [s, \frac{1+\lambda}{\lambda}s]$, $s \in S$, $\lambda > 0$. By the identification $T(U_0) = U_0 \times T_{x_0}(X)$ these vector fields are completely defined by their local representations

$$p_{\omega,s,\lambda} : U_0 \to T_{x_0}(X), \quad q_{s,\lambda} : U_0 \times [s, \frac{1+\lambda}{\lambda}s] \to T_{x_0}(X) ,$$

where U_0 is identified with an open set of $T_{x_0}(X)$.

Let J be an interval of $\mathbb{R}$ and denote by $\mathcal{D}^1(U_0, J)$ the set of bounded vector fields $f : U_0 \times J \to T_{x_0}(X)$ with the following properties:

(1) for each $t \in J$ the mapping $x \mapsto f(x, t)$ is of class C^1 on U_0 and the partial derivative with respect to the first variable $D_1 f$ is bounded on $U_0 \times J$;

(2) for any continuous mapping $\beta : J \to U_0$ the mappings $t \to f(\beta(t), t)$ and $t \to D_1 f(\beta(t), t)$ have one-sided limits at each point of J. The norm

$$||f|| = \sup_{(x,t) \in U_0 \times J} (||f(x, t)|| + ||D_1 f(x, t)||)$$

defines a Banach space structure on $\mathcal{D}^1(U_0, J)$. We denote by $\mathcal{D}^1(U_0)$ the closed subspace of vector fields which are independent of t.

Since X is locally compact, the neighborhood U_0 of x_0 may be chosen so that $p_{\omega,s,\lambda} \in \mathcal{D}^1(U_0)$, $q_{s,\lambda} \in \mathcal{D}^1(U_0, [s, \frac{1+\lambda}{\lambda}s])$ for any $\omega \in \Omega$, $s \in S$, $\lambda > 0$.

3. Throughout this section we assume that the given control $u : [0, t_u] \to \Omega$ steering x_0 to x_1 is not extremal, i.e. $K_0 = T_{x_0}(X)$.

Let $\dim X = k$ and V be an open ball in $T_{x_0}(X)$, of center 0 and radius $\rho < 1$. Choose k linearly independent elements $e_1, \ldots, e_k$ in V and consider the elements $v_j \in V$, $1 \leq j \leq 2k$, defined by $v_i = e_i$, $v_{k+i} = -e_i$ for $1 \leq i \leq k$. Take $c > \frac{1}{\| e_i \|}$ for all i; then each $v \in V$ may be written

$$v = \sum_{i=1}^{k} (c+a_i)e_i + \sum_{i=1}^{k} c(-e_i) = \sum_{j=1}^{2k} b_j v_j ,$$

where all $b_j > 0$. Since $V \subset K_0$ each v_j admits a representation as a finite linear combination of generators of the cone of attainability K_0 with positive coefficients.

Consequently, $v = \sum_{r=1}^{m} \lambda_r w_r$, where m and the w_r are <u>fixed</u>, $\lambda_r \geq 0$ and either

$$w_r = \alpha_*(s_r, 0).\xi(\alpha(s_r), \omega_r)$$

or

$$w_r = -\alpha_*(s_r, 0).\xi(\alpha(s_r), u(s_r)) ,$$

with $s_r \in S$, $\omega_r \in \Omega$. We may also assume, without loss of generality, that $s_1 \leq s_2 \leq \ldots \leq s_m$ and all $\lambda_r \leq \delta$ for some $\delta < 1$.

Let $I = [0, 1]$ and $C_m = \{c_0, c_1, \ldots, c_m\}$ a set of m+1 positive numbers such that $0 < c_0 < c_1 < \ldots < c_m < 1$, $c_1 - c_0 < s_1$ and $c_r - c_{r-1} < s_r - s_{r-1}$ if $s_r \neq s_{r-1}$ $(1 \leq r \leq m)$. We construct now for each $0 < \beta < 1$ a set $\mathcal{C}_\beta \subset \mathcal{D}^1(U_0, I)$.

For any $r (1 \leq r \leq m)$ and any $\theta_r (0 \leq \theta_r \leq \beta)$ define $h_r^{\theta_r} \in \mathcal{D}^1(U_0, I)$ as follows:

if $\theta_r = 0$, then $h_r^{\theta_r} = 0$ everywhere;

if $(x, \tau) \in U_0 \times ([0, c_{r-1}) \cup [c_r, 1])$, then $h_r^{\theta_r}(x, \tau) = 0$;

if $\theta_r \neq 0$ and $(x, \tau) \in U_0 \times [c_{r-1}, c_r)$,

then

$$h_r^{\theta_r}(x,\tau) = \begin{cases} p_{\omega_r,s_r,\theta_r}(x) & \text{for } w_r = \alpha_*(s_r,0).\xi(\alpha(s_r),\omega_r) \ , \\ q_{s_r,\theta_r}(x,\tau+s_r-c_{r-1}) & \text{for } w_r = -\alpha_*(s_r,0).\xi(\alpha(s_r),\ u(s_r)) \ . \end{cases}$$

For $\theta = (\theta_1, \ldots, \theta_m)$ take $h^\theta = \sum_{r=1}^{m} h_r^{\theta_r}$ and let $\mathcal{H}_\beta = \{h^\theta | 0 \leq \theta_r \leq \beta\}$.

Note that the set $\mathcal{H}_\beta \subset \mathcal{D}^1(U_0, I)$ depends not only on β , but also on the choice of the set C_m .

We shall establish now some properties of $\mathcal{H}_\beta$. Let $\mathcal{V}$ be a neighborhood of 0 in $\mathcal{D}^1(U_0, I)$ such that for each $f \in \mathcal{V}$ the integral curve α_f of the vector field f with initial condition $(x_0, 0)$ is defined on the entire interval I and lies in U_0 . Define $\Delta(f) = \alpha_f(1)$.

PROPOSITION 1. <u>The mapping</u> $\Delta : \mathcal{V} \to U_0$ <u>is of class</u> C^1 <u>and</u> $D\Delta(0).h = \int_0^1 h(x_0,\tau)d\tau$ <u>for</u> $h \in \mathcal{D}^1(U_0,I)$.

PROPOSITION 2. <u>For any</u> $0 < \beta < 1$ <u>we have</u>

$$(\mu_{0,t_u} \circ \Delta)(\mathcal{H}_\beta \cap \mathcal{V}) \subset A(x_0) \ .$$

PROPOSITION 3. <u>Let</u> $0 \leq \theta_r \leq \beta$ $(1 \leq r \leq m)$ <u>and</u> ε <u>be a real number such that</u> $0 \leq \varepsilon \leq 1$. <u>Then for every</u> $h^\theta \in \mathcal{H}_\beta$:

$$||h^{\varepsilon\theta}|| \leq \varepsilon||h^\theta|| \quad ,$$

<u>where</u> $\varepsilon\theta = (\varepsilon\theta_1, \ldots, \varepsilon\theta_m)$.

PROOF. Let $\tau \in [c_{r-1}, c_r)$. If $w_r = \alpha_*(s_r, 0) . \xi(\alpha(s_r), \omega_r)$ then obviously $||h^{\varepsilon\theta}(x, \tau)|| = \varepsilon||h^\theta(x, \tau)||$, $||D_1h^{\varepsilon\theta}(x, \tau)|| = \varepsilon||D_1h^\theta(x, \tau)||$ for all $x \in U_0$.

Suppose that $w_r = - \alpha_*(s_r, 0) . \xi(\alpha(s_r), u(s_r))$. Then the mapping $h^\chi \in \mathcal{H}_\beta$ takes on $U_0 \times [c_{r-1}, c_r)$ the same values as the local representation of the vector field

$$(x,\sigma) \mapsto -\chi(d_x\mu_{0,s_r})^{-1} . \xi(\mu_{0,s_r}(x), u(\sigma)) \text{ on } U_0 \times (s_r-\chi(c_r-c_{r-1}), s_r] \ .$$

Since $[s_r-\varepsilon\theta(c_r-c_{r-1}) , s_r] \subset [s_r-\theta(c_r-c_{r-1}), s_r]$ we get the required inequality.

PROPOSITION 4. <u>The set</u> $\bar{\mathcal{H}}_\beta$ <u>of</u> $\mathcal{D}^1(U_0, I)$ <u>is compact</u>.

PROOF. Consider the mappings $h_r^{\theta_r^+}, h_r^{\theta_r^-} \in \mathcal{D}^1(U_0, I)$ defined by

$$h_r^{\theta_r^+}(x,\tau) = \lim_{\substack{\chi_r \to \theta_r \\ \chi_r > \theta_r}} h_r^{\chi_r}(x,\tau),\quad h_r^{\theta_r^-}(x,\tau) = \lim_{\substack{\chi_r \to \theta_r \\ \chi_r < \theta_r}} h_r^{\chi_r}(x,\tau)$$

The existence of these limits follows from the properties of the vector field ξ_u. Note also that we may assume that $h_r^{\theta_r^+} = h_r^{\theta_r}$ for all $\theta_r < \beta$.

One shows first by a straightforward argument that

$$\bar{\mathcal{H}}_\beta = \{\bar{h}^\theta = \sum_{r=1}^m \bar{h}^{\theta_r} | 0 \le \theta_r \le \beta\},$$ where each $\bar{h}_r^{\theta_r}$ is either $h_r^{\theta_r}$ or $h_r^{\theta_r^-}$.

The compactness of $\bar{\mathcal{H}}_\beta$ follows then almost immediately.

PROPOSITION 5. <u>For</u> β <u>sufficiently small</u> $\bar{\mathcal{H}}_\beta \subset \mathcal{V}$ <u>and</u> $\Delta(\bar{\mathcal{H}}_\beta) = \Delta(\mathcal{H}_\beta)$.

PROOF. The first part of the statement follows easily from the definition of $\bar{\mathcal{H}}_\beta$ and from Proposition 3. Consider now for each $\bar{h}^\theta = \sum_{r=1}^m \bar{h}_r^{\theta_r} \in \bar{\mathcal{H}}_\beta$ the element $h^\theta = \sum_{r=1}^m h_r^{\theta_r} \in \mathcal{H}_\beta$. Then for every continuous mapping $\gamma : I \to U_o$ the mappings $\tau \mapsto \bar{h}^\theta(\gamma(\tau),\tau)$ and $\tau \longmapsto h^\theta(x(\tau),\tau)$ take different values at most countably many points. Hence the differential equations $x' = \bar{h}^\theta(x,\tau)$ and $x' = h^\theta(x,\tau)$ have the same solutions and thus $\Delta(\bar{h}^\theta) = \Delta(h^\theta)$.

We shall always assume that $\bar{\mathcal{H}}_\beta \subset \mathcal{V}$.

We define now a mapping $\Lambda : \mathcal{D}^1(U_0, I) \to T_{x_0}(X)$ by $\Lambda(h) = \sum_{r=1}^m (c_r - c_{r-1}) h(x_0, c_{r-1})$, where $c_0, c_1, \ldots, c_m$ are the (fixed) values used to define $\mathcal{H}_\beta$. Obviously Λ is linear and continuous.

PROPOSITION 6. <u>For every</u> $\varepsilon > 0$ <u>the set</u> $C_m = (c_0, \ldots, c_m)$ <u>in the definition of</u> $\mathcal{H}_\beta$ <u>may be chosen so that</u> $||\Lambda|| \le \varepsilon$ <u>and</u> $||D\Delta(0).h|| \le \varepsilon\, ||h||$ <u>for all</u> $h \in \bar{\mathcal{H}}_\beta$.

PROOF. The first statement is trivial. It is clearly sufficient to prove the second for $h \in \mathcal{H}_\beta$. Recall that $D\Delta(0).h = \int_0^1 h(x_0, \tau)d\tau$ by Proposition 1. Let

$h = h^\theta = \sum_{r=1}^{m} h_r^{\theta_r}$; then

$$||D\Delta(0).h^\theta|| \leq \sum_{r=1} ||D\Delta(0).h_r^{\theta_r}|| = \sum_{r=1}^{m} || \int_{c_{r-1}}^{c_r} h_r^{\theta_r}(x_0, \tau)d\tau|| \leq$$

$$\leq \sum_{r=1}^{m} (c_r - c_{r-1}) \sup_{\tau \in [c_{e-1}, c_r]} ||h_r^{\theta_r}(x_0, \tau)|| \leq ||h|| \sum_{r=1}^{m} (c_r - c_{r-1}) .$$

Consequently the required inequality holds for $c_r - c_{r-1}$ sufficiently small.

4. PROOF OF THE THEOREM. Assume that the control u is not extremal. It will be shown below that we can find a β, $0 < \beta < 1$, and two finite sets $\{s_1, \ldots, s_m\}$, $\{c_0, c_1, \ldots, c_m\}$ with the properties stated in the previous section such that $\mathcal{H}_\beta \subset \mathcal{V}$ and $x_0 \in \operatorname{Int} \Delta(\mathcal{H}_\beta)$. Then it follows from Proposition 2 that $\mu_{0,t_u}(\Delta(\mathcal{H}_\beta)) \subset A(x_0)$. Since μ_{0,t_u} is a diffeomorphism of U_0 we get $x_1 \in \operatorname{Int} \mu_{0,t_u}(\Delta(\mathcal{H}_\beta)) \subset \operatorname{Int} A(x_0)$, a contradiction.

It remains to prove that for a suitable choice of $C_m = \{c_0, c_1, \ldots, c_m\}$ and β we have $x_0 \in \operatorname{Int} \Delta(\mathcal{H}_\beta)$. Observe first that there exists a χ_0, $0 < \chi_0 < 1$, such that $\chi_0 V \subset \Lambda(\mathcal{H}_\beta)$, where V is the neighborhood of 0 in $T_{x_0}(X)$ introduced at the beginning of section 3. Indeed, if $v = \sum_{r=1}^{m} \lambda_r w_r \in V$, with $0 \leq \lambda_r \leq \delta$, take

$$\theta_r = \frac{\chi_0 \lambda_r}{c_r - c_{r-1}}, \quad \text{where} \quad \chi_0 \leq \frac{\beta}{\delta} \min_{1 \leq r \leq m} (c_r - c_{r-1}) .$$

Then $\theta_r \leq \beta$ and $\Lambda(h^\theta) = \sum_{r=1}^{m} (c_r - c_{r-1}) h_r^{\theta_r}(x_0, c_{r-1}) = \chi_0 v$.

Let now ρ be the radius of the ball V and choose C_m and β so that Proposition 6 holds for $\varepsilon = \frac{\rho}{8a}$, where a is the radius of $\mathcal{V}$. For every $v \in V$ consider an $h^\theta \in \mathcal{H}_\beta \subset \mathcal{V}$ such that $\Lambda(h^\theta) = \chi_0 v$. Then for any positive integer n :

$$\frac{\chi_0}{n} v = \Lambda(h^{\frac{\theta}{n}}) \quad .$$

By Proposition 3 $||h^{\frac{\theta}{n}}|| \leq \frac{1}{n}||h^{\theta}||$, hence $h^{\frac{\theta}{n}} \in \mathcal{H}_{\beta} \cap \frac{1}{n} \mathcal{V}$. Thus $\frac{\chi_0}{n} v \subset \Lambda(\bar{\mathcal{H}}_{\beta} \cap \frac{1}{n} \mathcal{V})$. Also

$$||D\Delta(0).h-\Lambda.h|| \leq ||D\Delta(0).h|| + ||\Lambda|| \, ||h|| \leq \frac{\rho}{4a}||h||$$

for all $h \in \bar{\mathcal{H}}_{\beta}$.

For every $L > 0$ there exists a $\sigma > 0$ such that for any pair $h^{\eta}, h^{\theta} \in \mathcal{H}_{\beta}$ with $\eta_r + \theta_r \leq \sigma$ $(1 \leq r \leq m)$:

$$||\Delta(h^{\eta+\theta}) - \Delta(h^{\theta}) - \Lambda.h^{\eta}|| \leq L \, ||h^{\eta}|| \quad .$$

The proof of this fact is rather lengthy and will be omitted here.

Define a subset $\mathcal{P}_{\beta}$ of $\mathcal{H}_{\beta}$ by : $h^{\theta} \in \mathcal{P}_{\beta}$ if and only if for every ε, $0 \leq \varepsilon < 1$, the inequality $||\Lambda(h^{\theta})|| \leq \varepsilon\rho$ implies $\theta_r \leq \varepsilon\beta$ for all $r (1 \leq r \leq m)$. The proof of the theorem is completed by applying the following analogue of the inverse mapping theorem to $Q = \bar{\mathcal{H}}_{\beta}$, $P = \mathcal{P}_{\beta}$, $u = \Lambda$, $f = \Delta$, $b = \rho$ and $\gamma_n = \frac{1}{n}$.

PROPOSITION 7. Let $\mathbb{E}, \mathbb{F}$ be Banach spaces, G an open set of $\mathbb{E}$ and $f : G \to \mathbb{F}$ a C^1-mapping. Let Q be a compact set in $\mathbb{E}$, $P \subset Q$ a subset such that $0 \in P \cap G$ and $u : \mathbb{E} \to \mathbb{F}$ a continuous linear mapping with the following properties:

(1) there exist open balls U_0 at 0 in $\mathbb{E}$ and V_0 at 0 in $\mathbb{F}$ of radii a and b respectively, a sequence of reals (γ_n), $\gamma_n > 0$, and a number $d > 0$ such that $\lim_{n\to\infty} \gamma_n = 0$, $\frac{\gamma_n}{\gamma_{n+1}} < d$ and $\gamma_n V_0 \subset u(P \cap \gamma_n U_0)$ for all n ,

(2) $||Df(0).x-u.x|| \leq \frac{b}{16a} ||x||$ for all $x \in Q \cap G$;

(3) <u>there exists an index</u> n_1 <u>such that for any</u> $n \geq n_1$ <u>and every finite sequence</u> $F = \{z_1, \ldots, z_k\} \subset P \cap \gamma_n \overline{U}_0$ <u>with</u> $\sum_{z \in F} ||u(z)|| \leq \gamma_n b$ <u>there is an</u> $x_F \in Q \cap \gamma_n \overline{U}_0$ <u>for which</u> : (a) $u(x_F) = \sum_{z \in F} u(z)$;

(b) <u>if</u> $z_0 \in P \cap \gamma_n \overline{U}_0$ <u>with</u> $||u(z_0)|| + \sum_{z \in F} ||u(z)|| \leq \gamma_n b$, then

$$||f(x_{F'}) - f(x_F) - u.z_0|| \leq \frac{b}{5ad} ||z_0||, \text{ where } F' = \{z_0, z_1, \ldots, z_k\} .$$

<u>In this case</u> $f(0) \in \text{Int } f(Q) \cap G)$.

References

1. F. ALBRECHT, Topics in Control Theory, Lecture Notes in Mathematics, Springer-Verlag, to appear.

2. H. HALKIN, On the Necessary Condition for Optimal Control of Nonlinear Systems, J. Anal. Math., <u>12</u>, 1-82 (1963).

3. E.B. LEE and L. MARKUS, Foundations of Optimal Control Theory, Wiley and Sons, 1967.

4. E. ROXIN, A Geometric Interpretation of Pontryagin's Maximum Principle, in Nonlinear Differential Equations and Nonlinear Mechanics, Academic Press, 1963.

SEMI-DYNAMICAL SYSTEMS *

Nam P. BHATIA

0. INTRODUCTION

I wish to take this opportunity to initiate a systematic study of what we shall, following HÁJEK [18], call Semi-Dynamical Systems.

Given a topological space X and the set of non-negative real numbers R^+, we define

(0.1) Definition.

If π is a map from $X \times R^+$ into X satisfying

(0.2) $\pi(x,0) = x$ for every $x \in X$,

(0.3) $\pi(\pi(x,t),s) = \pi(x,t+s)$ for every $x \in X$, and $t,s \in R^+$,

(0.4) π is continuous,

then the pair (X,π) is called a semi-dynamical system. We shall consistently write the image of (x,t) under π as $x\pi t$, so that (0.3) will read $(x\pi t)\pi s = x\pi(t+s)$.

Note that if in the above definition we replace R^+ by the set of real numbers R, we get the usual definition of a dynamical system, see e.g.[7].

It turns out that many of the notions, e.g., positive semitrajectory, positive limit set, positive prolongation, positive prolongational limit set, positive stability, positive attraction, positive asymptotic stability, so well known in Dynamical Systems, can immediately be introduced in this set-up, and we shall see that many theorems can be carried over with almost no change in proof. On the other hand, the notion of parallelizability looses its meaning, though some connected notions like dispersion can

---

* Partial support of the author by the National Science Foundation Grant No. NSF - GP - 7447 is gratefully acknowledged.

be carried over. Dual concepts with the adjective 'positive' replaced by 'negative' in the above statement are no longer available. One may introduce weak and strong negative concepts in the present set-up as we shall do, thus obtaining properties not relevant to the study of dynamical systems.

Indeed, the major reason for the introduction of the present study comes from the fact that functional-differential equations provide basic models for semi-dynamical systems as against differential equations which motivated the study of dynamical systems in the present form.

References [1-14] contain most known facts about dynamical systems. The remaining ones indicate that examples of semi-dynamical systems have been considered in the literature without necessarily labeling the mathematical object as a semi-dynamical system. HÁJEK [18] gives the concrete definition as given above.

The principle difference in the properties of a dynamical system and a semi-dynamical system perhaps occurs due to the fact that in the case of a dynamical system for any given $t \in R$, the map $\pi^t : X \to X$ defined by $\pi^t(x) = x\pi t$ is a homeomorphism of X onto X. This is not the case for similar maps defined by semi-dynamical systems.

We note that many recent developments in dynamical systems can be carried over with essentially no change in proofs in the case of a semi-dynamical system. We have therefore set before us the task of demonstrating some important properties of semi-dynamical systems which are well known for dynamical systems, and also to give a glimpse into the phenomena that may occur in the case of a semi-dynamical system, but are not known in the set-up of a dynamical system.

A detailed paper on semi-dynamical systems with complete proofs will appear elsewhere.

1. BASIC CONCEPTS

(1.1) Concepts carried over from dynamical systems:

Throughout these lectures X will denote a HAUSDORFF space unless otherwise specified. Given a semi-dynamical system (X,π) we define

(1.2) Definitions of positive trajectory, positive limit set, positive prolongation, and positive prolongational limit set.

For any $x \in X$ we set

(i) $\gamma^+(x) = \{x\pi t: t \in R^+\}$,

(ii) $\Lambda^+(x) = \bigcap\{\overline{\gamma^+(y)}: y \in \gamma^+(x)\}$,

(iii) $D^+(x) = \bigcap\{\overline{\gamma^+(U)}: U$ is a neighborhood of $x\}$,

(iv) $J^+(x) = \bigcap\{D^+(y): y \in \gamma^+(x)\}$.

The sets $\gamma^+(x)$, $\Lambda^+(x)$, $D^+(x)$, $J^+(x)$ are called the positive trajectory, the positive limit set, the positive prolongation, the positive prolongational limit set, respectively.

(1.3) Definition of positive invariance and positive minimality of a set.

Let M be a subset of X. We say that

(i) M is positively invariant if $\gamma^+(x) \subset M$ for every $x \in M$,

(ii) M is positively minimal if M is closed and positively invariant and no non-empty proper subset of M has these properties.

(1.4) Definition of stability, attraction, and asymptotic stability.

Let $M \subset X$ be a compact set. We say that

(i) M is positively stable if given any neighborhood U of M, there exists a neighborhood V of M such that $\gamma^+(V) \subset U$,

(ii) M is a positive weak attractor if the set

$A_\omega(M) \equiv \{x \in X: \Lambda^+(x) \cap M \neq \phi\}$ is a neighborhood of M,

(iii) M is a positive attractor if the set

$A(M) \equiv \{x \in X: \Lambda^+(x) \neq \phi$, and $\Lambda^+(x) \subset M\}$ is a neighborhood of M, and lastly,

(iv) M is positively asymptotically stable if M is positively stable and is a positive attractor.

Remark: For any $M \subset X$, the sets $A_\omega(M)$ and $A(M)$ introduced in (ii) and (iii) above are respectively called the region of positive weak attraction and the region of positive attraction of M.

(1.5) If $\phi : X \to 2^X$, where 2^X is the class of all subsets of X, and $M \subset X$, then we shall set $\phi(M) = \bigcup \{\phi(x) : x \in M\}$.

2. RESULTS ANALOGOUS TO RESULTS FOR DYNAMICAL SYSTEMS

(2.1) Properties of limit sets, prolongations, and prolongational limit sets.

(2.2) Theorem. Let $M \subset X$ be a compact positively invariant set. Then M contains a non-empty positively minimal set.

(2.3) Theorem. Given any $x \in X$

(i) $\overline{\gamma^+(x)} = \gamma^+(x) \cup \Lambda^+(x)$, and both $\overline{\gamma^+(x)}$ and $\Lambda^+(x)$ are closed and positively invariant sets.

(ii) If $\overline{\gamma^+(x)}$ is compact, or if $\Lambda^+(x)$ is compact and the space X is locally compact, then $\Lambda^+(x)$ is a compact and connected set.

(iii) $D^+(x) = \gamma^+(x) \cup J^+(x)$, and both $D^+(x)$ and $J^+(x)$ are closed and positively invariant sets.

(iv) If the space X is locally compact, then $D^+(x)$ and $J^+(x)$ are respectively connected whenever they are compact.

(v) If the space X is locally compact, then each component of the sets $\overline{\gamma^+(x)}$, $\Lambda^+(x)$, $D^+(x)$, $J^+(x)$ is non-compact, whenever the corresponding set is non-compact.

Remarks: Properties of $\Lambda^+(x)$, $J^+(x)$ as stated above can be improved considerably as will be seen in later sections.

(2.4) Theorem (Stability). Let X be locally compact, and $M \subset X$ be compact. Then M is positively stable if and only if $D^+(M) = M$.

(2.5) Theorem. Let $M \subset X$ be compact and x locally compact. Then M is positively asymptotically stable if and only if M is positively stable and M is a positive weak attractor.

(2.6) Lemma. If $M \subset X$ is compact and a positive weak attractor, then $A_\omega(M)$ is open and positively invariant. If M is a positive attractor, then $A_\omega(M) \equiv A(M)$.

(2.7) Theorem (Weak Attractors). If X is locally compact, and if $M \subset X$ is compact and a positive weak attractor, then $D^+(M)$ is compact and is the smallest positively asymptotically stable set containing M. Moreover, $A_\omega(M) \equiv A(D^+(M))$.

(2.8) Lemma. For any $x \in X$, and $\omega \in \Lambda^+(x)$, one has $J^+(x) \subset D^+(\omega)$.

3. LYAPUNOV FUNCTIONS AND ASYMPTOTIC STABILITY

In this section we assume that the space X is locally compact and metric with metric ρ.

(3.1) Theorem. Let $M \subset X$ be compact and asymptotically stable. Let $A(M)$ be the region of attraction of M. Then there exists a continuous functional $v: A(M) \to R^+$ having the following properties:

(i) $v(x) = 0$ for $x \in M$,

(ii) $v(x) > 0$ for $x \notin M$,

(iii) $v(x\pi t) < v(x)$ for $x \notin M$ and $t > 0$,

(iv) $v(x) \to +\infty$ as $x \to z \in \partial(A(M))$.

<u>Proof</u>: For $x \in A(M)$ define

$$\phi(x) = \sup\{\rho(x\pi t, M) : t \in R^+\} .$$

It is easily verified that $\phi(x)$ has the following properties:

(a) $\phi(x) = 0$ for $x \in M$,

(b) $\phi(x) > 0$ for $x \notin M$,

(c) $\phi(x\pi t) \leqq \phi(x)$ for $x \in A(M)$, $t \in R^+$,

and (d) $\phi(x)$ is continuous.

Now obtain $V : A(M) \to R^+$ by setting

$$V(x) = \int_0^\infty \phi(x\pi t) e^{-t} dt .$$

This V is easily verified to have the properties (a), (b), (c), and (d) of ϕ and in addition the property

$$(c') \quad V(x\pi t) < V(x) \quad \text{for } x \notin M, \; t > 0 .$$

This V need not have property (iv) required in the theorem. We now proceed as follows: Let $S[M,\epsilon] = \{x : \rho(x,M) \leqq \epsilon\}$ be a compact subset of $A(M)$. Let

$$m = \max\{V(x) : x \in S[M,\epsilon]\} .$$

Then $m > 0$. For any α, $0 < \alpha < m$, consider the set

$$B_\alpha = \{x : V(x) \leqq \alpha\} \cap S[M,\epsilon] .$$

It is now verified that B_α is a compact, positively invariant neighborhood of M and has the additional property that if $x \in \partial B_\alpha$, then $x\pi t \in \text{Int}(B_\alpha)$ for any $t > 0$. We now define the function $\tau : A(M) \to R^+$ as follows:

$$\tau(x) = \begin{cases} 0 & \text{for } x \in B_\alpha \\ \inf\{t > 0 : x\pi t \in B_\alpha\} & . \end{cases}$$

It is easily verified that τ is uniquely defined and is continuous. In fact, it is shown that $\tau(x) = t$ where $x\pi t \in \partial(B_\alpha)$ whenever $x \notin B_\alpha$. From this one obtains the property that if $x \notin B_\alpha$ for some $t \in R^+$, then

$$\tau(x\pi t) = \tau(x) - t .$$

We now define $v : A(M) \to R^+$ as follows:

$$v(x) = \begin{cases} V(x) & \text{if } x \in B_\alpha \\ \alpha e^{\tau(x)} & \text{if } x \notin B_\alpha \end{cases}$$

This $v(x)$ is claimed to have all the properties required by the theorems. To verify, e.g., (iv), let $\{x_n\}$ be a sequence in $A(M)$, $x_n \to x \in \partial(A(M))$. Assume that $\tau(x_n) \not\to \infty$. Then we may assume without loss of generality that $\tau(x_n) \to \tau \geqq 0$. But then by continuity of π, $x_n \pi \tau(x_n) \to x\pi t$, and as $x_n \pi \tau(x_n) \in \partial B_\alpha$, we conclude that $x\pi t \in \partial B_\alpha$, as ∂B_α is compact. But then $x \in A(M)$ a contradiction as $A(M)$ is open. This proves the theorem.

(3.2) <u>Theorem</u>. Let $M \subset X$ be a compact set. Let N be a neighborhood of M. Let $v : N \to R^+$ be a continuous functional having properties

(i) $v(x) = 0$ for $x \in M$,

(ii) $v(x) > 0$ for $x \notin M$,

(iii) $v(x\pi t) < v(x)$ if $x \notin M$, and $x\pi[0,t] \subset N$.

Then M is asymtotically stable.

<u>Remark</u>. It is not immediately evident that Theorem (3.1) can be improved to get results exactly analoguos to the one for dynamical systems, as for example given in [7] .

4. CONCEPTS PERTINENT TO SEMI-DYNAMICAL SYSTEMS

The main purpose of this section is to define the concept of a negative trajectory through a point $x \in X$. Since for any $x \in X$ and $t < 0$, $x\pi(-t)$ is meaningless where (X,π) is a semi-dynamical system, we proceed as follows:

(4.1) <u>Definition</u>. Given a semi-dynamical system (X,π) and $x \in X$, we set

$$F(x) = \{y \in X : x = y\pi t \quad \text{for some} \quad t \in R^+\}$$

and call the set $F(x)$ as the negative funnel through x. If $\gamma^-(x) \subset X$ has the properties

(i) $y, z \in \gamma^-(x) \Rightarrow y \in \gamma^+(z)$, or $z \in \gamma^+(y)$,

(ii) $\gamma^-(x) \subset F(x)$, and

(iii) $\gamma^-(x)$ is a maximal set with respect to properties (i) and (ii),

then $\gamma^-(x)$ is called a negative trajectory through x.

(4.2) <u>Definition</u>. A set $M \subset X$ will be called

(i) weakly negatively invariant if for each $x \in M$ there is at least one $\gamma^-(x) \subset M$,

(ii) negatively invariant if for each $x \in M$, one has $F(x) \subset M$.

(4.3) <u>Exercise</u>. M is negatively invariant if and only if $C(M)$ is positively invariant.

(4.4) <u>Definition</u>. $x \in X$ will be called a start point if for any $y \in X$ and $t > 0$, $y\pi t \neq x$.

The following theorem classifies negative trajectories through a point $x \in X$.

(4.5) <u>Theorem</u>. Let $\gamma^-(x)$ be a negative trajectory through x. Then one of the followings holds:

(i) there is a start point $y \in \gamma^-(x)$, in which case there is a $t > 0$ such that $y\pi t = x$ and $y\pi[0,t] = \gamma^-(x)$,

(ii) there is no start point in $\gamma^-(x)$, and $t_x \equiv \sup\{t > 0 : y \in \gamma^-(x), y\pi t = x\} < +\infty$,

and (iii) there is no start point in $\gamma^-(x)$, but the supremum in (ii) is $+\infty$.

(4.6) Remark. $\gamma^-(x)$ cannot contain more than one start point.

(4.7) Theorem. If $\gamma^-(x)$ is a negative trajectory containing no start points, and if $t_x < +\infty$, then any sequence $\{x_n\pi t_n: x_n \in \gamma^-(x),\ x_n\pi t_n \in \gamma^-(x),$ and $t_n \to t_x\}$ has the property that $x_n\pi t_n \to x$, but $\{x_n\}$ has no limit point in X.

Regarding the nature of positive limit sets and the prolongational limit sets, we can now make the following additional observation:

(4.8) Theorem. Let for any $x \in X$, $\overline{\gamma^+(x)}$ be compact. Then $\Lambda^+(x)$ is weakly negatively invariant, but contains no negative trajectories of the type (i) and (ii) in Theorem (4.5).

Proof: Let $y \in \Lambda^+(x)$. Then there is a sequence $\{t_n\}$, $t_n \to +\infty$, and $x\pi t_n \to y$. For any fixed $t > 0$, consider the sequences $\{t_n - t\}$ and $\{x\pi(t_n-t)\}$. The latter is defined for large n, is contained in the compact set $\overline{\gamma^+(x)}$, and $t_n - t \to +\infty$. We may therefore assume that $x\pi(t_n-t) \to z$. Then $z \in \Lambda^+(x)$, and since $x\pi t_n = (x\pi(t_n-t))\pi t$, we see on proceeding to the limit that $y = z\pi t$. We conclude that no point $y \in \Lambda^+(x)$ is a start point, and that for any $y \in \Lambda^+(x)$ there is a sequence of points $\{z_n\}$ $\Lambda^+(x)$ such that $z_n\pi 1 = z_{n-1}$, and $z_1\pi 1 = y$. This shows that there is a negative trajectory $\gamma^-(y)$ of type (iii) in Theorem (4.5) which is contained in $\Lambda^+(x)$. That no negative trajectory of type (ii) can lie in $\Lambda^+(x)$ follows from the fact that $\Lambda^+(x)$ is compact so that we will get a contradiction to Theorem (4.6) in such a case.

(4.9) Theorem. Let X be locally compact. Then for any $x \in X$, $\Lambda^+(x)$ and $J^+(x)$ are weakly negatively invariant, and contain no start points. If, moreover, $\Lambda^+(x)$ or $J^+(x)$ is compact, then in addition it does not contain any negative trajectory of type (ii).

Proof: We give only the proof for $J^+(x)$. The proof for $\Lambda^+(x)$ is similar. Let $y \in J^+(x)$. There are sequences $\{x_n\}$, $\{t_n\}$, $x_n \to x$, $t_n \to +\infty$, such that $x_n\pi t_n \to y$. For any fixed $t > 0$, consider for large n the sequences $\{x_n\pi(t_n-t)\}$, $\{t_n-t\}$, and let U be a compact neighborhood of y. Then $t_n - t \to +\infty$, and for large n, $x_n\pi t_n \in U$. Further, one of the following conditions holds: (a) $x_n\pi(t_n-t) \notin U$ for all sufficiently large n,

(b) there is a subsequence $\{x_{n_k}\pi(t_{n_k}-t)\} \subset U$. If (b) happens, we may assume that $x_{n_k}\pi(t_{n_k}-t) \to z \in J^+(x)$, but then $x_{n_k}\pi t_{n_k} = (x_n\pi(t_{n_k}-t)\pi t \to z\pi t$, so that $z\pi t = y$. If (a) happens, then there is a sequence $\{\tau_n\}$, $0 \leqq \tau \leqq t$, such that $x_n\pi(t_n-\tau_n) = (x_n\pi(t_n-t))\pi(t-\tau_n) \in \partial U$. We may assume then ($\partial U$ is compact) that $x_n\pi(t_n-\tau_n) \to z \in \partial U$, and $\tau_n \to \tau$, $0 \leqq \tau \leqq t$. Then as $x_n\pi t_n = (x_n\pi(t_n-\tau_n))\pi\tau_n$, we conclude by taking limits that $y = z\pi t$. Since $y \neq z$, t cannot be equal to zero. Further, $z \in J^+(x)$. Thus $J^+(x)$ contains no start points, and it does contain at least a part of a negative trajectory of any point $y \in J^+(x)$. It is now easy to see that the maximal extension of the segment of a negative trajectory obtained above must be a negative trajectory of either type (ii) or (iii) in Theorem (4.5).

We would like to close this section with a theorem on start points.

(4.10) Theorem. Let X be a locally compact space, and let (X,π) be a semi-dynamical system defined on X. Then the only isolated start points of X are those which have a neighborhood containing no points of X other than those on the positive trajectory through the start point.

Proof: Let $x \in X$ be an isolated start point, and let U be a compact neighborhood of x which contains no start points besides x. Let $\{x_n\} \subset U$ be a sequence with $x_n \to x$, $x_n \neq x$. Since no x_n is a start point, we can choose a negative trajectory $\gamma^-(x_n)$ for each n, such that $\gamma^-(x_n) \neq \{x_n\}$. There are now two possibilities: (a) there is a subsequence $\{x_{n_k}\}$ such that there are points $y_{n_k} \in \gamma^-(x_{n_k})$ with $\{y_{n_k}\} \subset U$, and $y_{n_k}\pi 1 = x_{n_k}$, $k = 1,2,\ldots$, or if (a) does not happen, then (b) there are sequences $\{y_n\}$, $\{\tau_n\}$, $y_n \in \gamma^-(x_n)$, $0 \leqq \tau_n \leqq 1$, $y_n\pi\tau_n = x_n$, and $y_n \in \partial U$ for $n = 1,2\ldots$. In case (a) we may assume $y_{n_k} \to z \in U$, and by continuity of π obtain $z\pi 1 = x$ contradicting that x is a start point. In case (b) we may assume that $y_n \to z \in \partial U$, and $\tau_n \to \tau$, $0 \leqq \tau \leqq 1$, so that continuity of π implies $z\pi\tau = x$. Since $z \neq x$ (note that U is a heighborhood of x and $z \in \partial U$), τ cannot be zero, so that x is a start point. But then $z\pi\tau = x$ contradicts the assumption that x is a start point. This proves the theorem.

(4.11) <u>Corollary</u>. If X is the euclidean n-space R^n, then the set of start points in R^n is either empty or uncountable.

(4.12) Conjecture. The set of start points in any euclidean space R^n, $n = 1,2...$, is empty.

5. SELF-INTERSECTING POSITIVE TRAJECTORIES

(5.1) <u>Definition.</u> Let (X,π) be a semi-dynamical system. A positive trajectory $\gamma^+(x)$, $x \in X$, will be called a self-intersecting trajectory if there exists $t_1, t_2 \in R^+$, $t_1 \neq t_2$, such that $x\pi t_1 = x\pi t_2$.

(5.2) <u>Definition</u>. A point $x \in X$ will be called a <u>positive rest point</u>, if $\gamma^+(x) \equiv \{x\}$.

(5.3) <u>Definition</u>. A point $x\ X$ will be called positively periodic if there is a $T > 0$ such that $x\pi t \equiv x\pi(t+T)$ for each $t \in R^+$.

(5.4) <u>Theorem</u>. If $x \in X$ is positively periodic, but is not a positive rest point, then there is a least $\tau > 0$ such that $x\pi(t+\tau) = x\pi\tau$ for each $t \in R^+$. Further, if $T > 0$ is a period of x, then $T = k\tau$ for some positive integer k.

<u>Proof</u>. Note that if $T > 0$ is a period of x, i.e. $x\pi t \equiv x\pi(t+T)$ for each $t \in R^+$, then so are nT for every positive integer n. Let now $\mathcal{P}$ denote the set of all positive periods of x. Set $\tau = g\ell b\, \mathcal{P}$. If $\tau = 0$, then there is a sequence $\{t_n\}$ of positive periods such that $t_n \to 0$. We claim that in this case x is a positive rest point. Note that if $T > 0$ is a period of x, then $\gamma^+(x) = x\pi[0,T]$. Thus, if x has a sequence of periods $\{\tau_n\}$, $\tau_n \to 0$, then $\gamma^+(x) = \bigcap_{n=1}^{\infty} x\pi[0,\tau_n]$. But the right-hand side is the set $\{x\}$, i.e., x is a rest point. Hence, $\tau > 0$. We need now show that τ is a period. Note, however, that if $\tau \notin \mathcal{P}$, then there is a sequence $\{\tau_n\} \subset \mathcal{P}$, $\tau_n \to \tau$ such that $x\pi t = x\pi(t+\tau_n)$ for each $t \in R^+$ and each τ_n. Then continuity of π implies that $x\pi t = x\pi(t+\tau)$ for each $t \in R^+$. Hence, τ is a period. Lastly, suppose that $T > 0$ is a period, but $T \neq k\tau$ for every positive integer k. Then there is a positive integer n such that $0 < t - n\tau < \tau$.

Clearly, $n\tau$ is a period, and we claim that $T - n\tau$ is also a period. This follows from the fact that for any $t \in R^+$,

$x\pi(t+T-n\tau) = (x\pi(t+T-n\tau))\pi n\tau = x\pi(t+T) = x\pi t$ as $n\tau$ and T are periods. But $T - n\tau < \tau$, contradicting the fact that τ was the least positive period. Hence, T must be of the form $k\tau$ for some positive integer k. This proves the theorem.

(5.5) Theorem. If $\gamma^+(x)$ is a self-intersecting trajectory, then $\Lambda^+(x)$ is the positive trajectory of a rest point or a periodic point.

Proof. Let $0 < t_1 < t_2$ be such that $x\pi t_1 = x\pi t_2$. Then we have, using axiom (0.3), that $x\pi(t_1+t) = x\pi(t_2+t)$ for every $t \in R^+$. Or, which is the same thing, $(x\pi t_1)\pi t = x\pi(t_1+t_2-t_1+t) = (x\pi t_1)\pi(t+t_2-t_1)$. This shows that the point $x\pi t_1$ is a periodic point, and $t_2 - t_1$ is a period of this point. We wish to show first that $x\pi t_1 \in \Lambda^+(x)$. This follows from the fact that if $T > 0$ is a period, then so is nT for every positive integer n. Thus we have

$x\pi t_1 = (x\pi t_1)\pi(n(t_2-t_1))$, $n = 1,2,\ldots$. Thus $x\pi t_1$ is the limit of a sequence $\{x\pi(t_1+n(t_2-t_1))\}$, where $t_1 + n(t_2-t_1) \to +\infty$ with n. This shows that $x\pi t_1 \in \Lambda^+(x)$. Lastly, we note that

$$\Lambda^+(x) \subset \overline{\gamma^+(x\pi t_1)} = \overline{(x\pi t_1)\pi[0,t_2-t_1]} = (x\pi t_1)\pi[0,t_2-t_1] ,$$

which proves the theorem.

(5.6) Corollary. If $\gamma^+(x)$ is a self-intersecting trajectory, then there is a $y \in \gamma^+(x)$ such that y is a rest point or a periodic point.

(5.7) Corollary. For any $x \in X$, $\gamma^+(x)$ is self-intersecting if and only if $\gamma^+(x)$ contains a rest point or a periodic point.

6. PROPERTIES OF COMPACT ASYMPTOTICALLY STABLE SETS

In this section we assume that X is a locally compact space and (X,π) is a semi-dynamical system defined on X.

(6.1) Theorem. Let $M \subset X$ be compact and let M and $X \setminus M$ be positively invariant. Further, let M be positively asymptotically stable with $A(M)$ as its region of attraction.

Then

(6.2) $$\partial M \setminus S \supset J^+(A(M) \setminus M)$$

where S denotes the set of all start points in X.

Proof. Let $x \in A(M) \setminus M$. Since $A(M)$ and $X \setminus M$ are positively invariant and open, $A(M) \setminus M = A(M) \cap (X \setminus M)$ is also positively invariant and open.

Thus, $J^+(x) \subset \overline{A(M) \setminus M} \subset \overline{X \setminus M}$. Now $x \in A(M) \Longrightarrow \Lambda^+(x) \cap M \neq \phi \Longrightarrow J^+(x) \subset D^+(\omega)$ where $\omega \in \Lambda^+(x) \cap M \Longrightarrow J^+(x) \subset D^+(M) = M$. We have thus proved that for any $x \in A(M) \setminus M$, $J^+(x) \subset M \cap \overline{X \setminus M} = \partial M$. Since $J^+(x)$ contains no start points (Theorem 4.9), we conclude that $J^+(A(M) \setminus M) \subset \partial M \setminus S$. The Theorem is proved.

(6.4) Theorem. Let (X,π) contain no start points. Then under the conditions of Theorem 6.3

$$\partial M = J^+(A(M) \setminus M).$$

Proof. We need only prove that $\partial M \subset J^+(A(M) \setminus M)$. Let $y \in \partial M$, and let $U \subset A(M)$ be a compact neighborhood of M. There is a sequence $\{y_n\} \subset A(M) \setminus M$ such that $y_n \to y$. Since there are no start points in $A(M)$, any negative trajectory $\gamma^-(y_n)$ through y_n is of the form (ii) or (iii) in Theorem 4.5. Let $\{\gamma^-(y_n)\}$ be any sequence of negative trajectories. We claim that $\gamma^-(y_n) \cap \partial U \neq \phi$ $n = 1,2,\ldots$. This happens for if $\gamma^-(y_n)$ is of type (ii) in Theorem 4.5, then it must leave every compact set by Theorem 4.6, and if $\gamma^-(y_n)$ is of type (iii) and does not intersect ∂U, then we set

$$L = \left\{ x \mid \exists \{z_k\} \subset \gamma^-(y_n),\ \{t_k\} \subset R^+,\ z_k \pi t_k = y_n,\ t_k \to \infty,\ z_k \to x \right\}.$$

Since $\gamma^-(y_n) \subset U$, L is not empty, and we claim it is positively invariant. This last assertion follows from the fact, that if $x \in L$ and $\{z_k\}$, $\{t_k\}$ are sequences with $t_k \to +\infty$, $z_k \to x$, $z_k \pi t_k = y_n$, then for any $t > 0$, consider the sequence $z_k \pi t \to x\pi t$. Indeed $z_k \pi t \in \gamma^-(y_n)$ for all large k, and also $(z_k \pi t)\pi(t_k - t) = z_k \pi t_k = y_n$ for all large k. Moreover, $t_k - t \to +\infty$. Hence, $x\pi t \in L$ and L is positively invariant. L is moreover closed as may easily be verified. Now $L \subset U \subset A(M)$. If $L \cap M \neq \phi$, then we have a sequence $\{z_k\} \subset \gamma^-(y_n)$, $z_k \to z \in L \cap M$, and $z_k \pi t_k = y_n$, and $y_n \notin M$, so that M cannot be stable. If $L \cap M = \phi$, then the fact that L is a non-empty compact positively invariant subset of $A(M)$ leads to the conclusion that there is an $x \in L \subset A(M)$ such that $\phi \neq \Lambda^+(x) \subset L$ and $\Lambda^+(x) \cap M = \phi$, which is absurd as M is asymptotic stable. Thus, each negative trajectory $\gamma^-(y_n)$ intersects ∂U . Let now for each n , $x_n \in \gamma^-(y_n) \cap \partial U$. Then there is a sequence $\{t_n\} \subset R^+$ such that $x_n \pi t_n = y_n$. We may assume that $x_n \to x \in \partial U$, as ∂U is compact. Then we have $x_n \to x \in \partial U$, $x_n \pi t_n = y_n \to y \in \partial M$. But then t_n must tend to $+\infty$ for otherwise we may assume that $t_n \to t \in R^+$, so that $y = x\pi t$, and as $X \setminus M$ is positively invariant $y \notin M$, a contradiction. This shows, however, that $y \in J^+(x)$ for some $x \in A(M) \setminus M$. The proof is completed.

(6.5) Lemma. Let $M \subset X$ be compact. Let M contain a sequence of positively periodic points $\{x_n\}$, with the sequence of corresponding positive periods $\{T_n\}$ such that $T_n \to 0$. The M contains a positive rest point.

(6.6) Theorem. Let $M \subset R^n = X$ be a compact set which is globally weakly attracting. Then M contains a positive rest point.

Concluding Remarks. It will be interesting to find which properties of compact minimal sets of dynamical systems can be carried over to compact positively minimal sets of semi-dynamical systems and what, if any, properties do they possess.

Acknowledgments. (i) The author is indebted to Professor Ottomar HAJEK for many productive and useful discussion of the material presented above.

(ii) Professor H. HALKIN supplied to us a proof of our conjecture (4.12) after the lectures in Varenna. We gratefully Thank Prof. HALKIN for the same and from now on would like to label this conjecture as a Theorem.

References

[1] H. POINCARE, "Les méthodes nouvelles de la mecaniques celeste", Gauthier-Villars, Paris (1892-1899). Reprint, Dover, New York.

[2] Ivar BENDIXON, Sur les courbes definies par des equations differentielles, Acta Mathematica, 24 (19o1), 1-88.

[3] A.M. LYAPUNOV, "Probleme géneral de la stabilite du mouvement", Annals of Mathematics Studies, No. 17, Princeton, 1947.

[4] G.D. BIRKHOFF, "Dynamical Systems", American Mathematical Society Colloquium Publications, Vol. 9 New York, 1927.

[5] V.V. NEMYTSKII and V.V. STEPANOV, "Qualitative Theory of Differential Equations", Princeton University Press, Princeton, 1960. Original Russian 1947, 1949.

[6] V.V. NEMYTSKII, Topological Problems of the Theory of Dynamical Systems, Uspehi Mat. Nauk., 4 (1949), 91-153 (Russian); English Translation: Am. Math. Soc. Translations, No. 1o3, (1954).

[7] N.P. BHATIA and G.P. SZEGÖ, "Dynamical Systems: Stability Theory and Applications", Lecture Notes in Mathematics, No. 35, Springer-Verlag, Berlin-Heidelberg-New York, 1967.

[8] V.I. ZUBOV, "Methods of A.M. LYAPUNOV and Their Application", Noordhoff, Groningen, The Netherlands, 1964; Original Russian, 1957.

[9] T. URA, Sur le courant exterieur a unse region invariante; Prolongements d'une caracteristique et l'ordre de stabilite, Funkcialaj Ekvacioj, Vol. 2 (1959), 143-200.

[10] T. URA, On the flow outside a closed invariant set; stability, relative stability and saddle sets, Contributions to Differential Equations, 3 (1964), 249-294.

[11] J. AUSLANDER and P. SEIBERT, Prolongations and Stability in Dynamical Systems, Annales de l'Institut Fourier, 14 (1964), 237-267.

[12] N.P. BHATIA, On Asymptotic Stability in Dynamical Systems, Mathematical Systems Theory, 1 (1967, 113-127.

[13] N.P. BHATIA, Weak Attractors in Dynamical Systems, Bol. Soc. Mat. Mex., 11,(1966), 56-64.

[14] N.P. BHATIA and G.P. SZEGÖ, Weak Attractors in R^n, Mathematical Systems Theory, 1 (1967), 129-133.

[15] N.N. KRASOVSKII, "Stability of Motion", Stanford Univ. Press, Stanford, 1963; Original Russian, 1959.

[16] J.K. HALE, A Class of Functional-Differential Equations, Contributions to Differential Equations, 1, (1963), 411-423.

[17] J.K. HALE, Sufficient Conditions for Stability and Instability of Autonomous Functional-Differential Equations, J. of Differential Equations, 1 (1965), 452-482.

[18] O. HÁJEK, Structure of Dynamical Systems, Comm. Math. Univ. Carolinae, 6 (1965), 53-72. Correction same journal, 6 (1965), 211-212.

[19] O. HÀJEK, Critical Points of Abstract Dynamical Systems, Comm. Math., Univ. Carolinae, 5 (1964), 121-124.

[20] F. BROCK FULLER, On the surface of section and periodic trajectories, Amer. J. Math., 87 (1965), 473-480.

- - - - - - - - - - - - -

Correction. (i) The set $J^+(x)$ as introduced in 1.2(iv) does not have in general the property claimed in Theorem 4.9. However if $J^+(x)$ is defined by

$$J^+(x) = \{y: \exists \{x_n\}, \{t_n\}, x_n \to x, t_n \to +\infty \text{ and } x_n \pi t_n \to y\},$$

then this set has the desired property. For dynamical systems the two definitions are equivalent, that this is not the case in semi-dynamical systems can be shown on examples.

(ii) The definition of a negative trajectory in 4.1 does not yield the classification in Theorem 2.6. This discrepancy can be removed by replacing 4.1(i) by: 4.1(i)' if $y, z \epsilon \gamma^-(x)$ and $t_y = \inf\{t \geqq 0: y\pi t_y = x\}$, $t_z = \inf\{t > 0: z\pi t_z = x\}$, then either there is a τ_y, $0 \leqq \tau_y \leqq t_y$, such that $y\tau_y = z$ or there is a τ_z, $0 \leqq \tau_z \leqq t_z$, such that $z\tau_z = y$. The definition then reads: For any $x \epsilon X$, a set $\gamma^-(x) \subset X$ will be called a negative semi-trajectory through x whenever either $\gamma^-(x) \equiv \{x\}$ and x is a rest point, or $\gamma^-(x)$ satisfies conditions 4.1(i), (ii), and (iii).

For details see the author's forthcoming paper aon 'Local Semi-Dynamical Systems' in joint authorship with O. **HÀJEK.**

SOME THEOREMS IN MEASURE THEORY AND GENERALIZED DYNAMICAL SYSTEMS DEFINED BY CONTINGENT EQUATIONS (*)

Charles Castaing

I

In this section, we establish two theorems related to the theory of relaxed control introduced by Gamkrelidze, Ghouila-Houri, Warga, and others. We first give some notations and definitions.

Let T and Z be two compact spaces. Denote by $\mathcal{C}(T)$ (resp. $\mathcal{C}(Z)$) the Banach spaces of real continuous functions defined on T (resp. Z). Let $\mathcal{M}(T)$ (resp. $\mathcal{M}(Z)$) be the dual of $\mathcal{C}(T)$ (resp. $\mathcal{C}(Z)$) equipped with the weak star topology $\sigma(\mathcal{M}(T),\mathcal{C}(T))$ (resp. $\sigma(\mathcal{M}(Z),\mathcal{C}(Z))$ and let $\mathcal{M}^+(T)$ (resp. $\mathcal{M}^+(Z)$) be the cone of positive measures on T (resp. Z). Let π be a continuous mapping of Z into T. For every $\nu \in \mathcal{M}^+(Z)$ define:

$$\int f \, d\pi(\nu) = \int f\circ\pi \, d\nu$$

for all $f \in \mathcal{C}(T)$. The measure $\pi(\nu)$ is called the image of measure ν under the continuous mapping π. Observe that the mapping $\nu \dashrightarrow \pi(\nu)$ of $\mathcal{M}^+(Z)$ into $\mathcal{M}^+(T)$ is continuous.

Let now T be a compact space, μ a positive Radon measure on T and U be a compact space. Denote by π the projection of T×U on T. Let

$$Q = \{\sigma \in \mathcal{M}^+(T\times U) : \pi(\sigma) = \mu\} .$$

Then Q is compact with respect to the weak star topology. Denote by $\mathcal{F}_0$ the set of simple μ-measurable mapping of T into U. Let $\mathcal{F}$ be the set of μ-measurable mapping of T into U. For every μ-measurable mappings v of T into U, let:

$$\sigma_v = \int_T \delta_{(t,v(t))} \, d\mu(t) \quad ,$$

where $\delta_{(t,v(t))}$ is the Dirac mass at (t,v(t)). It can be verified that $\sigma_v \in Q$. Denote

---

(*) This work has been achieved under the direction of Prof. Pallu de La Barriere with the support of DGRT (convention de Recherches 63 FR 199)

by $Q_{\mathcal{F}_0}$ (resp. $Q_{\mathcal{F}}$) the set of positive measures $(\sigma_v;\ v\in\mathcal{F}_0)$ (resp. $(\sigma_v;\ v\in\mathcal{F})$. We have the following:

THEOREM 1. - Suppose that T is metrisable and μ is atomless. Then $Q_{\mathcal{F}_0}$ is dense in Q with respect to the weak star topology σ ((T×U),$\mathcal{C}$(T×U)).

Proof. - Let $\sigma \in Q$. Then there exists by virtue of Ionescu-Tulcea's theorem ([14],Th.3, p. 461) a scalarly μ-measurable mapping $t \dashrightarrow \lambda_t$ of T into the set $\mathcal{M}^1(U)$ of probability measures on U such that

$$\langle\sigma,f\rangle = \int_T \langle\delta_t \otimes \lambda_t,\ f\rangle\ d\mu(t)$$

for all $f \in \mathcal{C}(T\times U)$. Let ϵ be a positive number and (f_k); $k = 1,\dots,m$, a finite sequence of elements of $\mathcal{C}(T\times U)$. We must find a positive measure σ_v defined by

$$\sigma_v = \int_T \delta_{(t,v(t))} d\mu(t), \quad (v \in \mathcal{F}_0)$$

such that:

(1) $$|\langle\sigma,f_k\rangle - \langle\sigma_v,f_k\rangle| \leq \epsilon, \quad k = 1,2,\dots,m.$$

We first observe that $\mathcal{C}(T) \otimes \mathcal{C}(U)$ identified to a subspace of $\mathcal{C}$(T×U is dense in this Banach space. Hence for proving (1), we can obviously choose the elements f_k $(k=1,\dots,m)$ in $\mathcal{C}(T) \otimes \mathcal{C}(U)$; and the proof will be achieved if we show the following:

Let $\epsilon > 0$, (g_i) and (h_i); $i = 1,2,\dots,n$, two finite sequences of elements in $\mathcal{C}(T)$ and $\mathcal{C}(U)$ respectively. Then there exists a positive measure $\sigma_v = \int_T \delta(t,v(t))d\mu(t)$, $(v \in \mathcal{F})$ such that:

$$|\langle\sigma,g_i \otimes h_i\rangle - \langle\sigma_v,g_i \otimes h_i\rangle| \leq \epsilon\ , \quad i = 1,2,\dots,n.$$

Put $M = \sup_{1\leq i\leq n} \int_T |g_i(t)|\ d\mu(t)$. Consider a finite covering of U by open sets (Ω_j); $j = 1,\dots,p$ such that:

$$u,u' \in \Omega_j \Longrightarrow |h_i(u) - h_i(u')| \leq \frac{\epsilon}{M}\ ; \quad i = 1,2,\dots,n.$$

Let (φ_j), $j = 1,\dots,p$ be a partition of unity subordinated to the covering (Ω_j). Observe first that if $u_j \in$ support (φ_j), then we have:

(2) $$|h_i(u) - h_i(u_j)|\ \varphi_j(u) \leq \frac{\epsilon}{M}\ \varphi_j(u)$$

for all $u \in U$; $i=1,2,\dots,n$, $j=1,2,\dots,p$.

According to (2) we deduce that:

$$(3) \qquad |h_i(u) - \sum_{j=1}^{p} \varphi_j(u)\, h_i(u_j)| \leq \frac{\varepsilon}{M}$$

for all $u \in U$ and $i=1,2,...n$. Now define:

$$\dot{\lambda}_t = \sum_{j=1}^{p} \alpha_j(t)\delta_{u_j} \qquad (t \in T)$$

where $\alpha_j(t) = \langle\lambda_t,\varphi_j\rangle$, $(t \in T)$ and u_j support $\in (\varphi_j)$. Then $t \dashrightarrow \dot{\lambda}_t$ is a scalarly μ-measurable mapping of T into $\mathcal{M}^1(U)$. Let:

$$\dot{\sigma} = \int_T \delta_t \otimes \lambda_t \, d\mu(t) .$$

We have:

$$(4) \qquad \langle\dot{\sigma},g_i \otimes h_i\rangle = \int_T g_i(t) \langle\dot{\lambda}_t,h_i\rangle \, d\mu(t)$$

$$= \int_T \sum_{j=1}^{p} \alpha_j(t)\; g_i(t)\; h_i(u_j)\; d\mu(t) .$$

Moreover we have:

$$\langle \sigma,g_i \otimes h_i\rangle - \langle\dot{\sigma},g_i \otimes h_i\rangle = \int_T g_i(t) \langle\lambda_t-\dot{\lambda}_t,h_i\rangle \, d\mu(t)$$

for $i = 1,2,...,n$. Using (3) we obtain

$$(5) \qquad |\langle \sigma,g_i \otimes h_i\rangle - \langle\dot{\sigma},g_i \otimes h_i\rangle| \leq \varepsilon , \quad i = 1,2,...,n.$$

According to Ljapunov's theorem (See |11|) there exists a μ-measurable mapping $t \dashrightarrow (\dot{\alpha}_1(t),...,\dot{\alpha}_p(t))$ taking its values in the profile of the simplex Λ_p such that:

$$\int_T \alpha_j(t) g_i(t)\, d\mu(t) = \int_T \dot{\alpha}_j(t) g_i(t)\, d\mu(t)$$

for $i = 1,2,...,n$; $j = 1,2,...,p$.

It follows that:

$$\langle\dot{\sigma},g_i \otimes h_i\rangle = \int_T \sum_{j=1}^{p} \dot{\alpha}_j(t) g_i(t) h_i(u_j)\, d\mu(t); \quad i = 1,2,...,n.$$

Setting $T_j = \{t \in T : \dot{\alpha}_j(t) = 1\}$; $j = 1,2,...,p$. The sets T_j are μ-measurable and determine a finite partition of T. Define $v(t) = u_j$ for $t \in T_j$; $j = 1,2,...,p$. Then $v \in \mathcal{F}_0$. Let $\sigma_v = \int \delta_{(t,v(t))} d\mu(t)$.

We have:

$$\langle \dot{\sigma}, g_i \otimes h_i \rangle = \sum_{j=1}^{p} \int_{T_j} g_i(t)\, h_i(u_j)\, d\mu(t)$$

$$= \langle \sigma_v, g_i \otimes h_i \rangle \; ; \quad i = 1,2,\ldots,n.$$

Using (5) we obtain:

$$|\langle \sigma, g_i \otimes h_i \rangle - \langle \sigma_v, g_i \otimes h_i \rangle| \leq \epsilon \; ; \; i = 1,2,\ldots,n.$$

This completes the proof.

In the following theorem, suppose that U is <u>metrisable</u>. Let E be a real separated locally convex quasi complete space. Suppose there exists in the dual E' of E a countable $\sigma(E',E)$ dense subset. Let h be a continuous mapping of $T \times U$ into E.

THEOREM 2. - <u>Let</u>

$$A = \{h\, d\sigma \, : \, \sigma \in Q\}$$
$$B = \{h\, d\sigma_v \, : \, v \in \mathfrak{F}\}$$
$$C = \{h\, d\sigma_v \, : \, v \in \mathfrak{F}_o\}$$

<u>Then:</u>

a) A <u>is a convex compact subset of</u> E.

b) <u>If</u> $h(t,U)$ <u>is convex for all</u> $t \in T$, <u>we have</u>: $A = B$.

c) <u>If</u> μ <u>is atomless, we have</u>: $\overline{C} = \overline{B} = A$.

Proof. - a) We first observe that $\int h\, d\sigma \in E$ ([1], Int., ch.3, p. 80). Moreover, the mapping $\sigma \dashrightarrow \int h\, d\sigma$ is continuous on Q(equipped with the weak star topology). It follows that A is convex and compact because Q is convex and compact.

b) Obviously the set $Q_{\mathfrak{F}}$ of positive measure (σ_v; $v \in \mathfrak{F}$) is included in Q. Then we must show that $A \subset B$. Let σ be an element of Q. By virtue of ([14], Th.3, p. 461) there exists a scalarly μ-measurable mapping $t \dashrightarrow \lambda_t$ of T into the set $\mathcal{M}^1(U)$ of probability measures on U such that

$$\sigma = \int_T \delta_t \otimes \lambda_t \, d\mu(t).$$

We have:

$$\int_{T\times U} h(t,u)\ d\sigma(t,u) = \int_T d\mu(t) \int_U h(t,u)\ d\lambda_t(u).$$

Let $g(t) = \int_U h(t,u)\ d\lambda_t(u)$ for all $t\in T$. Then $g(t) \in h(t,U)$ for all $t\in T$ because $h(t,U)$ is convex for all $t\in T$ by hypothesis. Now we claim that the mapping $t \dashrightarrow g(t)$ is μ-measurable. In fact it is clear the the scalar function $t \dashrightarrow \langle e', g(t)\rangle = \int \langle e', h(t,u)\rangle\ d\lambda_t(u)$ is μ-measurable for all $e' \in E'$. Therefore g is μ-measurable by virtue of ([4], Th.6, p. 4). Now apply ([3] Cor. 5.1) to pick a μ-measurable mapping v of T into U such that $g(t) = h(t,v(t)) = \langle h, \delta_{(t,v(t))}\rangle$ for all $t\in T$. Then we have:

$$\int h\ d\sigma = \int h\ d\sigma_v .$$

c) According to theorem 1, the set of positive measures $(\sigma_v;\ v\in\mathcal{F}_0)$ is dense in Q equipped with the weak star topology. Recall that the mapping $\sigma \dashrightarrow \int h\ d\sigma$ is continuous, then we have:

$$\overline{C} = \overline{B} = A.$$

<u>An application to control theory.</u>

Let U be a compact metrisable space, and let f be a continuous mapping of $(0,T)\times R^n\times U$ into R^n. An <u>admissible control</u> is a measurable mapping u of (0,T) into R^n such that

$$\frac{dX(t)}{dt} = f(t,X(t),u(t)) \quad \text{a.e. in } (0,T).$$

Let: $F(t,x)$ the convex hull of the set $\{f(t,x,v) : v\in U\}$. A <u>relaxed admissible trajectory</u> is an absolutely continuous mapping X of (0,T) into R^n such that

$$\frac{dX(t)}{dt} \in F(t,X(t)) \quad \text{a.e. in } (0,T).$$

THEOREM 3. - <u>Let</u> X <u>be a relaxed admissible trajectory. Then there exists a positive measure</u> σ <u>defined by the formula</u>

$$\sigma = \int_0^T \delta_t \otimes \lambda_t\ dt$$

<u>where</u> $t \dashrightarrow \lambda_t$ <u>is a scalarly</u> dt-<u>measurable mapping of</u> (0,T) <u>into the set</u> ${}^1(U)$ <u>of probability measures on</u> U <u>such that:</u>

$$X'(t) = \int_U f(t,X(t),v)\, d\lambda_t(v);\ t \in (0,T) \quad \text{a.e.}$$

Proof. - Put

$$h(t,\mu) = \int_U f(t,X(t),v)\, d\mu$$

for $t\in(0,T)$ and $\mu\in\mathcal{M}^+(U)$. Then $t \dashrightarrow h(t,\mu)$ is continuous on $(0,T)$ for every $\mu\in\mathcal{M}^+(U)$, and $\mu \dashrightarrow h(t,\mu)$ is continuous on $\mathcal{M}^+(U)$ for every $t\in(0,T)$. But $\mathcal{M}^1(U)$ is compact metrisable, then we can apply the corollary 5.2' of theorem 5.2. (See 3); taking into account of the fact that $X'(t) \in h(t,\mathcal{M}^1(U))$ a.e. in t. Therefore there exists a scalarly dt-measurable mapping $t \dashrightarrow \lambda_t$ of $(0,T)$ into $\mathcal{M}^1(U)$ such that:

$$\lambda_t \in \mathcal{M}^1(U) \text{ and } X'(t) = h(t,\lambda_t) \quad \text{a.e.}$$

This completes the proof.

II

Let us consider now the multivalued differential equation

(I) $$\frac{dx}{dt} \in F(t,x),\ t\in R,\ x\in\Omega \text{ (open} \neq \emptyset) \text{ in } R^n.$$

Suppose that the following conditions are satisfied:

1) $F(t,x)$ is convex compact $\neq \emptyset$, $\forall t \in R, \forall x \in \Omega$,

2) $x \dashrightarrow F(t,x)$ is upper semi continuous on $\Omega, \forall t \in R$,

3) $t \dashrightarrow F(t,x)$ is measurable with respect to the Lebesgue measure on $R, \forall x \in \Omega$,

4) there exists a locally integrable function g on R such that

$$\|u\| \leq g(t),\ \forall u \in F(t,x),\ \forall t \in R, \forall x \in \Omega.$$

Definition. - We call solution of differential equation (I) an absolutely continuous mapping of X of an interval (a,b) into Ω such that

$$\frac{dX(t)}{dt} \in F(t,X(t)) \quad \text{a.e.}$$

Let now M be a convex compact non void subset $\subset \Omega$. Let T_o be a positive number such that

$$\int_o^{T_o} g(s)\, ds \leq d(M, \Omega)$$

and let $T \in (0, T_o)$. Denote by $\mathcal{C}_{R^n}(0,T)$ the Banach space of continuous mappings of $(0,T)$ into R^n.

THEOREM 4. - *For every* $\xi \in M$, *there exists at least one solution* X *of equation (I) defined on* $(0,T)$ *such that* $X(0) = \xi$. *The set* S_ξ *of solutions* X *such that* $X(0) = \xi$ *is compact and non void in* $\mathcal{C}_{R^n}(0,T)$; *moreover, the compact-valued mapping* $\xi \dashrightarrow S_\xi$ *of* M *into* $\mathcal{C}_{R^n}(0,T)$ *is upper semi-continuous on* M.

The proof of this theorem is omitted and can be found in the author's paper "Sur les équations différentielles multivoques" [2].

Let us consider now the multivalued differential equation

$$\text{(II)} \qquad \frac{dx}{dt} \in F(t,x), \quad t \in R, \quad x \in \Omega \text{ (open} \neq \emptyset) \subset R^n.$$

Suppose that the following conditions are satisfied:

1) $F(t,x)$ is convex compact $\neq \emptyset$, $\forall t \in R$, $\forall x \in \Omega$,
2) $x \dashrightarrow F(t,x)$ is upper semi-continuous on Ω, $\forall t \in R$,
3) $t \dashrightarrow F(t,x)$ is measurable with respect to the Lebesgue measure on R, $x \in \Omega$,
4) there exists an upper semi-continuous function φ on Ω such that:

$$\|u\| \leq \varphi(x), \quad u \in F(t,x), \; \forall t \in R, \; \forall x \in \Omega .$$

Remark. - If F is a compact-valued upper semi-continuous mapping of Ω into R^n, then the condition 4) is automatically satisfied. In fact, the function

$$\varphi : x \dashrightarrow \max \{ u \; : u \in F(x)\}, \; x \in \Omega$$

is upper semi-continuous.

Now we make use of a following method due to GHOUILA-HOURI.
Let M be a convex compact $\neq \emptyset$ subset in Ω. Let $r_o = d(M, \Omega)$. For every $\rho \in (0, r_o)$ let:

$$M(\rho) = \max \{\varphi(x) : d(x,M) \leq \rho\}.$$

Define $T_o = \int_0^{r_o} \frac{d\rho}{M(\rho)}$. For every $t \in (0,T_o)$ we define the function $t \dashrightarrow r(t)$ by the formula $\int_0^{r(t)} \frac{d\rho}{M(\rho)} = t$. We have:

$$\frac{dr(t)}{dt} = M(r(t)) \text{ for } t \in (0,T_o).$$

Let $T \in (0,T_o)$.

THEOREM 5. - For every $\xi \in M$, there exists at least one solution X of equation (II) defined on $(0,T)$ such that $X(0) = \xi$. The set S_ξ of solutions X such that $X(0) = \xi$ is compact and non void in the Banach space $\mathcal{C}_{R^n}(0,T)$; moreover, the compact-valued $\xi \dashrightarrow S_\xi$ of M into $\mathcal{C}_{R^n}(0,T)$ is upper semi-continuous on M.

We employ the same argument as in the proof of theorem 4:
Denote by X_M the set of absolutely continuous mapping of $(0,T)$ into R^n such that $X(0) \in M$ and that $\frac{dX(t)}{dt} \leq M(r(t))$ a.e.. Then X_M is convex and compact and $X(t) \in \Omega, \forall t \in (0,T), \quad X \in X_M$.

Now we can reproduce word for word the same argument used in the proof of theorem 4.

1 BOURBAKI, N.: Intégration, Livre VI, ch. 4.

2 CASTAING, Ch.: Sur les équations différentielles multivoques, C.R., t. 263, p. 63-66, 1966

3 ——————— Sur les multi-applications mesurables, Faculté des sciences Caen 1966

4 ——————— Sur un théoreme de représentation intégrale lié a la comparaison des mesures, Faculté des Sciences Caen 1967

5 CESARI, L.: Existence theorem for optimal solution in Lagrange and Pontrjagin problems, J SIAM CONTROL, vol. 3, 1966

6 FILIPPOV, A.F.: On certain questions in the theory of optimal control, Vesnik Moskov, Ser. Math. 1959

7 ——————— Differential equations with many valued discontinuous right hand side Doklady Akad. Nauk, SSSR, t. 151, 1963

8 GAMKRELIDZE, R.V.: Optimal sliding states, Doklady, t. 143, no. 2

9 GHOUILA-HOURI: Problemes d'existence de commandes optimales, Seminaire d'automatique théorique Faculté des Sciences Caen 1965

0 ——————— Equations différentielles multivoques, C.R. t. 261, 1965, p. 2568-2570

1 KARLIN, S.: Extreme points of vector functions, Proc. Amer. Math. Soc. Vol. 4, p. 603-610, 1953

2 LASOTA, A. OPIAL, Z.: An application of the Kakutani Ky-Fan theorem in the theory of ordinary differential equation, Bul.Ac.Pol.Sc.Math., vol. XIII, no. 11-12, 1965

3 PLIS, A.: Measurable orientor fields, Bul.Acad.Pol.Sc.Math., vol. 13, no. 8, 1965

4 IONESCU-TULCEA, A. IONESCU-TULCEA, C.: Disintegration of measures, Ann. INST. FOURIER, Grenoble, t. 14, no. 2, 1964, p. 445-472

5 WARGA, J.: Relaxed variational problems, Journal of Math. Analysis and Applications, vol. 4, 1962, p. 111-128

6 WAZEWSKI, T.: Sur une généralisation de la notion des solutions d'une équation au contingent, Bull. Acad. Pol. Math. t. 10, p. 11-15

On the controllability of linear difference-differential systems

A. Halanay

In this note we shall obtain some new properties related with the concept of complete controllability, following the ideas exposed in the Appendix A of the book of V.M. Popov [1].

Consider a control system with time lag of the form

$$\dot{x}(t) = Ax(t) + Bx(t-\tau) + b\eta(t) \qquad (1)$$

where $B = bc^*$, hence the columns of the matrix B are colinear with the vector b. Since the system is completely defined by the matrix A and vectors b,c we shall denote it by (A,b,c).

For a motivation in considering this class of systems remark that if an equation of n-th order

$$x^{(n)}(t) = \sum_{k=1}^{n} \alpha_k x^{(k-1)}(t) + \sum_{k=1}^{n} \beta_k x^{(k-1)}(t-\tau) + \mu(t) \qquad (2)$$

is written in usual way as a system we get

$$y(t) = \tilde{A}y(t) + \tilde{B}y(t-\tau) + \tilde{b}\mu(t)$$

$$\tilde{A} = \begin{pmatrix} 0 & 1 & 0 \dots 0 \\ 0 & 0 & 1 \dots 0 \\ \dots & \dots & \dots \\ 0 & 0 & 0 \dots 1 \\ \alpha_1 & \alpha_2 & \alpha_3 \dots \alpha_n \end{pmatrix}, \quad \tilde{B} = \begin{pmatrix} 0 & 0 \dots 0 \\ 0 & 0 \dots 0 \\ \dots & \dots \\ 0 & 0 \dots 0 \\ \beta_1 & \beta_2 \dots \beta_n \end{pmatrix}, \quad \tilde{b} = \begin{pmatrix} 0 \\ 0 \\ \cdot \\ 0 \\ 1 \end{pmatrix} \qquad (3)$$

hence the colums of the matrix $\tilde{B}$ are colinear with the vector $\tilde{b}$.

<u>Definition.</u> <u>The system</u> (1) <u>is completely controllable if for every</u> $T > 0$ <u>and for every continuous vector-function</u> φ <u>defined on</u> $[-\tau, 0]$ <u>there esixts a control</u> u, <u>piece-wise continuous and such that the solution of the system corresponding to</u>

the initial function φ and the control u is vanishing on $[T,T+\tau]$.

Remark. The condition $B = bc^*$ implies that for the complete controllability it suffices that the solution equals zero for $t = T$, hence if $x(T) = 0$ we can choose the control u for $t > T$ defined by $u(t) = -c^*x(t-\tau)$ and for $t > T$ the system will reduce to $\dot{x}(t) = A(t)x(t)$, $x(T) = 0$, hence $x(t) \equiv 0$ for $t > T$.

Proposition. System (1) is completely controllable if and only if (A,b) is completely controllable, (i.e., if the system $\frac{dx}{dt} = Ax + bu$ is completely controllable).

Proof. a) If (1) is completely controllable, by taking $0 < T < \tau$ we deduce that for every initial function φ there is a control u such that for the solution of the system $\dot{x}(t) = Ax(t) + bc^*\varphi(t-\tau) + bu(t)$, $x(0) = \varphi(0)$ we shall have $x(T) = 0$; hence there is a control ν such that for the solution of the system $\dot{x}(t) = Ax(t) + b\nu(t)$, $x(0) = \varphi(0)$ we shall have $x(T) = 0$, $(\nu(t) = u(t) + c^*\varphi(t-\tau))$. It follows that (A,b) is completely controllable.

b) Let (A,b) be completely controllable; for $0 < T < \tau$ there exists a control ν such that for the solution of $\dot{x}(t) = Ax(t) + b\nu(t)$, $x(0) = \varphi(0)$ we have $x(T) = 0$; choose $u(t) = \nu(t)-c^*\varphi(t-\tau)$ and for the control u the solution of system (1) with the initial function φ will vanish for $t = T$. Let now $T > \tau$, k such that $(k-1)\tau \leq T < k\tau$; there exists ν such that for the solution of $\dot{y}(t) = Ay(t) + b\nu(t)$, $y((k-1)\tau) = x((k-1)\tau)$ we have $y(T) = 0$. We may choose $u(t) = \nu(t) - c^*x(t-\tau)$ for $(k-1)\tau \leq t < T$ and $u(t)$ arbitrary for $[0,(k-1)\tau]$.

Corollary. An equation of the form (2) is always completely controllable.

Proposition 2. The class of systems considered is invariant under linear transformations.

Proof. Let $y = Sx$; we obtain

$$\dot{y}(t) = \tilde{A}y(t) + \tilde{b}\tilde{c}^*y(t-\tau) + \tilde{b}u(t) , \quad \tilde{A} = SAS^{-1}, \quad \tilde{b} = Sb , \quad \tilde{c} = (S^{-1})^*c$$

Corollary. System (1) is completely controllable if and only if it is linear equivalent to an equation (2).

Proof. a) If the system is completely controllable, (A,b) is completely controllable, hence there exists a non singular matrix S such that $SAS^{-1} = \tilde{A}$, $Sb = \tilde{b}$, where $\tilde{A}, \tilde{b}$ have the form in (3). From $\tilde{B} = \tilde{b}\tilde{c}^*$ it follows that $\tilde{B}$ has also the form in (3).

b) If the system is linear equivalent to the equation (2) we have $\tilde{A} = SAS^{-1}$, $\tilde{b} = Sb$, $\tilde{A}, \tilde{b}$ as in (3) and (A,b) is completely controllable.

Definition. The characteristic function of system (1) is

$$\chi(\sigma) = \det(\sigma E - A - e^{-\sigma\tau} B) .$$

We see that

$$\chi(\sigma) = \sigma^n - \sum_{k=1}^{n} (a_k + b_k e^{-\sigma\tau})\sigma^{k-1} .$$

Proposition 3. Suppose the systems (A,b,c), $(\hat{A},\hat{b},\hat{c})$ have the same characteristic function. If they are completely controllable then they are linear equivalent.

Proof. Let $\det(\sigma E - A - Be^{-\sigma\tau}) = \det(\sigma E - \hat{A} - \hat{B}e^{-\sigma\tau}) = \sigma^n - \sum_{k=1}^{n} (\alpha_k + \beta_k e^{-\sigma\tau})\sigma^{k-1}$.

It follows that $\det(\sigma E - A) = \det(\sigma E - \hat{A}) = \sigma^n - \sum_{k=1}^{n} \alpha_k \sigma^{k-1}$; (A,b) and $(\hat{A},\hat{b})$ are completely controllable, hence there exist S_1 and S_2 such that $S_1 A S_1^{-1} = \tilde{A}$, $S_2 \hat{A} S_2^{-1} = \tilde{A}$, $S_1 b = \tilde{b}$, $S_2 \hat{b} = \tilde{b}$, where $\tilde{A}, \tilde{b}$ are as in (3). We deduce that $S_2^{-1} S_1 A S_1^{-1} S_2 = S_2^{-1} \tilde{A} S_2 = \hat{A}$, $S_2^{-1} S_1 b = S_2^{-1} \tilde{b} = \hat{b}$, $c^* S_1^{-1} S_2 = \tilde{c}^* S_2 = \hat{c}^*$ where $\tilde{c}^* = (\beta_1, \ldots, \beta_n)$.

Proposition 4. Let x_0 be the solution of (1) such that $x_0(t) \equiv 0$ for $t < 0$, $x_0(0) = b$, and let $W(t) = \int_0^t x_0(s) x_0^*(s) ds$. System (1) is completely controllable if and only if $W(t) > 0$ for all $t > 0$.

Proof. a) If (1) is completely controllable, then (A,b) is completely controllable and $q^* e^{At} b = 0$ implies $q = 0$. Let $0 < t \leq \tau$, and q arbitrary; we have

$$q^*W(t)q = \int_0^t q^*x_0(s)x_0^*(s)q\,ds = \int_0^t |q^*x_0(s)|^2 ds \geq 0 .$$

If $q^*W(t)q = 0$ it follows that $q^*x_0(s) \equiv 0$, $s \in [0,t]$; but for $s \in [0,\tau]$ we have $x_0(s) = e^{As}b$ hence $q = 0$. We have thus $W(t) > 0$ for $0 < t \leq \tau$ and since $q^*W(t)q$ is increasing it follows that $W(t) > 0$ for all $t > 0$.

b) Let $W(T) > 0$, hence $W(T)$ has an inverse. For given φ we choose $u(t) = x_0^*(T-t)W^{-1}(T)u$,

$$u = -X(T)\varphi(0) - \int_{-\tau}^{0} x_0(T-\alpha-\tau)c^*\varphi(\alpha)d\alpha$$

where $X(t)$ is the matrix solution of the system

$$\dot{x}(t) = Ax(t) + Bx(t-\tau)$$

defined by $X(0) = E$, $X(t) \equiv 0$ for $t < 0$.

We have for the corresponding solution of (1)

$$x(t) = X(t)\varphi(0) + \int_{-\tau}^{0} X(t-\alpha-\tau)bc^*\varphi(\alpha)d\alpha + \int_0^t X(t-\alpha)bu(\alpha)d\alpha$$

and since $X(t)b = x_0(t)$ we get

$$x(t) = X(t)\varphi(0) + \int_{-\tau}^{0} x_0(t-\alpha-\tau)c^*\varphi(\alpha)d\alpha + \int_0^t x_0(t-\alpha)u(\alpha)d\alpha$$

and

$$x(T) = X(T)\varphi(0) + \int_{-\tau}^{0} x_0(T-\alpha-\tau)c^*\varphi(\alpha)d\alpha + \int_0^T x_0(T-\alpha)x_0^*(T-\alpha)W^{-1}(T)u\,d\alpha =$$

$$= -u + \int_0^T x_0(\beta)x_0^*(\beta)d\beta\, W^{-1}(T)u = 0 .$$

<u>Corollary 1.</u> <u>If there exists</u> $T > 0$ <u>such that</u> $q^*x_0(t) \equiv 0$ <u>for</u> $t \in [0,T]$ <u>implies</u> $q = 0$ <u>the system is completely controllable.</u> We have indeed $q^*W(T)q = \int_0^T |q^*x_0(s)|^2 ds = 0$; if $q^*W(T)q = 0$ it follows $q^*x_0(s) = 0$ for $s \in [0,T]$ hence $q = 0$ and $W(T) > 0$ and for every initial state there exists a control u transferring it to the origin in the time T. If the system was not completely controllable, (A,b) were not completely controllable and by a linear transformation we should get

$$SAS^{-1} = \begin{pmatrix} A_{11} & A_{12} \\ 0 & A_{22} \end{pmatrix} \qquad SB = \begin{pmatrix} b_1 \\ 0 \end{pmatrix} \qquad SBS^{-1} = \begin{pmatrix} b_1 \\ 0 \end{pmatrix} (c_1^* c_2^*) = \begin{pmatrix} b_1c_1^* & b_1c_2^* \\ 0 & 0 \end{pmatrix}$$

hence the system

$$\dot{y}_1(t) = A_{11}y_1(t) + A_{12}y_2(t) + b_1c_1^*y_1(t-\tau) + b_1c_2^*y_2(t-\tau) + b_1u(t)$$

$$\dot{y}_2(t) = A_{22}y_2(t) .$$

If the control μ transfers x from the state φ to the origin in time T, then it transfers y from $S\varphi$ to the origin, and if φ is arbitrary $S\varphi$ is also arbitrary; but if $(S\varphi)(0)$ has the last coordinate different from zero, the solution y can never vanish. It follows that (A,b) is completely controllable.

If (1) is completely controllable then for all $T > 0$ from $q^*x_0(t) \equiv 0$ on $[0,T]$ it follows $q = 0$ since from $q^*x_0(t) \equiv 0$ on $[0,T]$ it follows that $q^*e^{At}b \equiv 0$.

2. Let y_q^T the solution of the adjoint system

$$\dot{y}(t) = -y(t)A - y(t+\tau)B$$

defined on $t < T$ by the conditions $y(T) = q^*$, $y(t) \equiv 0$ for $t > T$. If (1) is completely controllable then from $y_q^T(t)b \equiv 0$ on $[t_0,T]$ it follows $q = 0$ hence $y_q^T \equiv 0$. We have indeed $y_q^T(t) = q^*X(T-t)$ and $y_q^T(t)b \equiv 0$ means that $q^*X(T-t)b \equiv 0$ on $[t_0,T]$ hence $q^*x_0(t) \equiv 0$ on $[0,T-t_0]$ and $q = 0$.

If there is a $T > 0$ and $t_0 < T$ such that from $y_q^T(t)b \equiv 0$ on $[t_0,T]$ follows $y_q^T \equiv 0$ the system (1) is completely controllable, since this condition means that $q^*x_0(t) \equiv 0$ on $[0,T-t_0]$ implies $q = 0$.

Proposition 5. The following statements are equivalent:

1) (A,b) is completely controllable

2) For every function $\chi(\sigma) = \sigma^n - \sum_{k=1}^{n} (\gamma_k + \delta_k e^{-\sigma\tau})\sigma^{k-1}$

there exist q_1, q_2 such that

$$\det(\sigma E - A - bq_1^* - (B + bq_2^*)e^{-\sigma\tau}) \equiv \chi(\sigma)$$

and q_1, q_2 are uniquely determined.

3) If $\pi(\sigma)$ is an arbitrary polynomial of degree $n-1$ there exists q such that

$$q^*(\sigma E-A-Be^{-\sigma\tau})^{-1}b = \frac{\pi(\sigma)}{\det(\sigma E-A-Be^{-\sigma\tau})}$$

and q *is uniquely determined.*

4) *If* $q^*(\sigma E-A-Be^{-\sigma\tau})^{-1}b \equiv 0$ *then* $q = 0$.

5) *For all distinct complex numbers* σ_j $(j = 1,2,\dots,n)$ *such that* $\det(\sigma_j E-A-Be^{-\sigma_j\tau}) \neq 0$ *the vectors* $(\sigma_j E-A-Be^{-\sigma_j\tau})^{-1}b$ *are linearly independent.*

6) *There exists* σ_j $(j = 1,2,\dots,n)$ *such that the vectors* $(\sigma_j E-A-Be^{-\sigma_j\tau})^{-1}b$ *are linearly independent.*

Proof. a) If (A,b) is completely controllable, system (1) is linear equivalent to an equation (2) for which the characteristic function is $\sigma^n - \sum_{k=1}^{n}(\alpha_k + \beta_k e^{-\sigma\tau})\sigma^{k-1}$; if we take $\tilde{q}_1^* = (\gamma_1 - \alpha_1,\dots,\gamma_n - \alpha_n)$, $\tilde{q}_2^* = (\delta_1 - \beta_1,\dots,\delta_n - \beta_n)$ we get $\det(\sigma E-\tilde{A}-\tilde{b}\tilde{q}_1^* - (\tilde{B}+\tilde{b}\tilde{q}_2^*)e^{-\sigma\tau}) \equiv \sigma^n - \sum_{k=1}^{n}(\gamma_k+\delta_k e^{-\sigma\tau})\sigma^{k-1} =$ where $\tilde{A},\tilde{B},\tilde{b}$ are as in (3). Let S be the matrix of this linear transformation, $q_1 = S^*\tilde{q}_1$, $q_2 = S^*\tilde{q}_2$. We have $\det(\sigma E-A-bq_1^*-(B+bq_2^*)e^{-\sigma\tau}) = \det(\sigma E-A-b\tilde{q}_1^*S-(B+b\tilde{q}_2^*S)e^{-\sigma\tau})$ $\det(\sigma E-\tilde{A}-\tilde{b}\tilde{q}_1^*-(\tilde{B}+\tilde{b}\tilde{q}_2^*)e^{-\sigma\tau}) \equiv \chi(\sigma)$. The unicity of q_1 and q_2 follows from the unicity of $\tilde{q}_1$ and $\tilde{q}_2$.

If we have the property 2), then $\delta_k = 0$, $k = 1,\dots,n$, $q_2 = -c$, we deduce that there exists q_1 such that $\det(\sigma E-A-bq_1^*) = \sigma^n - \sum_{k=1}^{n}\gamma_k\sigma^{k-1}$, hence (A,b) is completely controllable.

Hence statements 1) and 2) are equivalent.

If (A,b) is completely controllable then system (1) is linear equivalent to an equation (2); for this equation we have

$$(\sigma E-\tilde{A}-\tilde{B}e^{-\sigma\tau})\begin{pmatrix}1\\ \sigma\\ \vdots\\ \sigma^{n-1}\end{pmatrix} = \det(\sigma E-\tilde{A}-\tilde{B}e^{-\sigma\tau})\tilde{b} \quad ,$$

hence

$$\tilde{q}^*(\sigma E-\tilde{A}-\tilde{B}e^{-\sigma\tau})^{-1}\tilde{b} = \frac{\pi(\sigma)}{\det(\sigma E-\tilde{A}-\tilde{B}e^{-\sigma\tau})} \quad ,$$

where $\pi(\sigma)$ is the given polynomial and $\tilde{q}$ is uniquely defined by the coefficients of $\pi(\sigma)$. If $q = S^*\tilde{q}$ we get $q^*(\sigma E-A-Be^{-\sigma\tau})^{-1}b = \dfrac{\pi(\sigma)}{\det(\sigma E-A-Be^{-\sigma\tau})}$, hence 1) implies 3).

The unicity of q in 3) shows that 3) implies 4). If we have 4) then $b \neq 0$ and we can choose a basis such that b will be the last vector of this basis. If S is the matrix the columns of which are the vectors of this basis we have $S^*b = \begin{pmatrix} 0 \\ \vdots \\ |b|^2 \end{pmatrix}$.

By the linear transformation defined by S^* we get a system for which $\hat{B}$ has the first rows equal to zero and $(\sigma E-\hat{A}-\hat{B}e^{-\sigma\tau})^{-1}\hat{b}$ is colinear to the last column of $(\sigma E-\hat{A}-\hat{B}e^{-\sigma\tau})^{-1}$; the elements of this column are the algebraic complements of the elements of the last row of $\sigma E-\hat{A}-\hat{B}e^{-\sigma\tau}$ divided by $\det(\sigma E-\hat{A}-\hat{B}e^{-\sigma\tau})$. But the elements of the last row of the matrix $\sigma E-\hat{A}-\hat{B}e^{-\sigma\tau}$ are the only ones containing $e^{-\sigma\tau}$ hence their algebraic complements are polynomials in σ of degree less or equal to $n-1$. It follows that for all $\hat{q}$ we have $\hat{q}^*(\sigma E-\hat{A}-\hat{B}e^{-\sigma\tau})^{-1}\hat{b} = \dfrac{\pi(\sigma)}{\det(\sigma E-\hat{A}-\hat{B}e^{-\sigma\tau})}$, where π is a polynomial of degree at most $n-1$. Let $q = S\hat{q}$; from $\hat{b} = S^*b$, $\hat{A} = S^*A(S^*)^{-1}$, $\hat{B} = S^*B(S^*)^{-1}$ it follows that $\dfrac{\pi(\sigma)}{\det(\sigma E-A-Be^{-\sigma\tau})} = q^*(\sigma E-A-Be^{-\sigma\tau})^{-1}b$.

If we have 4) and 5) were not true, then for distinct σ_j $(\sigma_j E-A-Be^{-\sigma_j\tau})^{-1}b$ were linear dependent and thus for a $q \neq 0$ $q^*(\sigma_j E-A-Be^{-\sigma_j\tau})^{-1}b = 0$ hence $\pi(\sigma_j) = 0$ for π a polynomial of degree $n-1$. That implies $\pi(\sigma) \equiv 0$ and by 4) $q = 0$. We get a contradiction hence 4) implies 5). That 5) implies 6) is obvious. We shall see that 6) implies that (A,b) is completely controllable and the proposition will be proved. If (A,b) were not completely controllable, then for a non singular S

$$SAS^{-1} = \hat{A} = \begin{pmatrix} A_{11} & A_{12} \\ 0 & A_{22} \end{pmatrix}, \quad Sb = \hat{b} = \begin{pmatrix} b_1 \\ 0 \end{pmatrix}, \quad SBS^{-1} = \hat{B} = \begin{pmatrix} B_{11} & B_{12} \\ 0 & 0 \end{pmatrix}$$

$(\sigma_j E-\hat{A}-\hat{B}e^{-\sigma_j\tau})^{-1}\hat{b} = \begin{pmatrix} F_{11} & b_1 \\ 0 & \to 0 \end{pmatrix}$ and these vectors could not be linear independent; from $(\sigma_j E-\hat{A}-\hat{B}e^{-\sigma_j\tau})^{-1}\hat{b} = S(\sigma_j E-A-Be^{-\sigma_j\tau})^{-1}b$ we see that $(\sigma_j E-A-Be^{-\sigma_j\tau})^{-1}b$ could not be linear independent.

Reference

[1] V.M. Popov Hiperstabilitatea sistemelor automate
Editura Academiei Republicii Socialiste România 1966.

THE CORE AND COMPETITIVE EQUILIBRIA

Werner Hildenbrand

There is at least a formal connection between Control Theory and Mathematical Economics. Some recent important results in equilibrium analysis of perfectly competitive economies (i.e., economies with an atomless measure space of economic agents)[1] are based on mathematical theorems which are equally basic mathematical tools in Control Theory.

The theorems we have in mind are:

Theorem of Liapunov [6][2]

Let μ denote an R^n-valued atomless measure defined on a σ-algebra $\mathcal{A}$. Then, the range of μ, i.e., the set $\{\mu(E) | E \in \mathcal{A}\}$ is a convex and compact subset in R^n.

Theorem on Measurable Selections

Let $(A, \mathcal{A}, \nu)$ be a finite measure space and ϕ a correspondence of A into R^n such that the graph G_ϕ of ϕ belongs to the product σ-algebra $\mathcal{A} \otimes \mathcal{B}(R^n)$.[3] Then there exists an $\mathcal{A}$-measurable function f of A into R^n such that $f(a) \in \phi(a)$ for ν-almost every $a \in A$.

This version of a Measurable Selection Theorem is a generalization of a lemma of VON NEUMANN [7] (Lemma 5, p. 448) due to R. J. AUMANN [3]. An alternative approach to prove Measurable Selection Theorems is given in M. SION [8]. The above theorem still holds when the range space R^n is an arbitrary Souslin space (i.e., a metrizable topological space which is a continuous image of a complete separable metric space).

---

[1] We follow the notation and terminology of G. Debreu's Lectures presented in this volume. The reader not familiar with economic equilibrium analysis is referred to these lectures.

[2] A short and easy (but not elementary) proof of Liapunov's Theorem has been given by J. LINDENSTRAUSS, J. Math. and Mech., 15, 971-972, 1966.

[3] $\mathcal{B}(S)$ denotes the σ-algebra of Borel subsets of the topological space S.

The purpose of this lecture is to show how the above theorems are used in Mathematical Economics to prove a theorem on the core of an economy with an atomless measure space of economic agents. For a detailed introduction and a more explicit proof see W. HILDENBRAND [5].

2. Let $S = R^{\ell}$ denote a finite-dimensional commodity space[4] and let $(A, \mathcal{A}, \nu)$ be a normalized measure space.

Each element a in the set A represents an economic agent whose characteristics are:

his <u>resources</u>,[4] given by a point ω_a in the commodity space S,
his <u>consumption set</u>,[4] given by a nonempty subset X_a in S,
his <u>preferences</u>,[4] given by an irreflexive binary relation $<_a$ on X_a,
his <u>production set</u>,[4] given by a subset Y_a in S with $0 \in Y_a$.

<u>DEFINITION</u>: A normalized measure space $(A, \mathcal{A}, \nu)$, a resource distribution ω, a consumption set correspondence X, a preference function $<$ and a production set correspondence Y defines an <u>economy</u> $\mathcal{E}$ <u>with a measure space of economic agents if</u>

(I) the function ω of A into S is ν-integrable,

(II) the graph of the correspondence X of A into S and the graph of the preference function $<$ are measurable, i.e.,
$\{(a,x) \in A \times S \mid x \in X_a\} \in \mathcal{A} \otimes \mathcal{B}(S)$ and
$\{(a,x,z) \in A \times S \times S \mid x <_a z\} \in \mathcal{A} \otimes \mathcal{B}(S) \otimes \mathcal{B}(S)$,

(III) the graph of the correspondence Y of A into S is measurable, i.e.,
$\{(a,y) \in A \times S \mid y \in Y_a\} \in \mathcal{A} \otimes \mathcal{B}(S)$.

Assumptions (I) - (III) are purely technical and have no economic interpretation.

The measurability of the correspondence Y implies the following result:

<u>Let</u> p <u>be any vector in</u> S, <u>then the function</u> $a \to \sup_{y \in Y_a} (p|y)$ <u>of</u> A <u>into</u> $R \cup (\infty)$ <u>is nonnegative and</u> ν-<u>measurable (hence quasi-integrable) and for every</u> $E \in \mathcal{A}$ <u>holds</u>:

---

[4] This concept is defined in G. DEBREU [4] resp. in G. Debreu's lectures presented in this volume.

$$\sup\{(p|\int_E Y d\nu)\} = \int_E \sup(p|Y)d\nu,$$ [5]

i.e., given the price system p, the supremum profit for the coalition E is equal to the sum of the supremum profits of the individual members of the coalition E.

The measurability follows from the identity

$$\{a \in A | \sup(p|Y_a) > \lambda\} = \text{proj}_A\{(a,y) \in A \times S | y \in Y_a \text{ and } (p|y) > \lambda\}$$

and the following well-known theorem: $M \in \mathcal{A} \otimes \mathcal{B}(S)$ implies $\text{proj } M \in \text{Souslin}[\mathcal{A}] \subset \mathcal{A}_\nu$, where $\mathcal{A}_\nu$ denotes the completion of the σ-algebra $\mathcal{A}$ with respect to ν.

Using the Measurable Selection Theorem the proof of the above equation is straightforward.

An allocation in the economy $\mathcal{E}$ is a ν-integrable function x of A into S such that ν-almost everywhere $x_a \in X_a$.

An allocation x is called attainable (resp. attainable for the coalition $E \in \mathcal{A}$) if $\int_A x d\nu \in \int_A \omega d\nu + \int_A Y d\nu$ (resp. $\int_E x d\nu \in \int_E \omega d\nu + \int_E Y d\nu$).

An allocation x is said to be blocked by the coalition $E \in \mathcal{A}$ if there exists an allocation z such that

(i) $x_a <_a z_a$ for almost every $a \in E$,

(ii) $\nu(E) > 0$ and z is attainable for the coalition E.

The core of the economy $\mathcal{E}$ is the set of all unblocked attainable allocations.

An attainable allocation x is said to be competitive, if there is a price system $p \in S$, $p \neq 0$, such that

(C.1) for almost every $a \in A$, x_a is a maximal element with respect to $<_a$ in the budget set $\gamma_a(p) = \{z \in X_a | (p|z) \leqq (p|\omega_a) + \sup_{y \in Y_a}(p|y)\}$

[5] The integral of a correspondence ϕ of a measure space $(F, \mathcal{F}, \mu)$ into R^n is defined by the set $\int_F \phi d\mu = \{\int_F f d\mu | f \in \mathcal{L}_\phi\}$, where $\mathcal{L}_\phi$ denotes the set of all μ-integrable functions f of F into R^n such that $f(a) \in \phi(a)$ almost everywhere in F.

(C.2) $(p|\int_A(x-\omega)d\nu) = \max_{y\in\int_A Yd\nu} (p|y)$, i.e., the total production $\int_A(x-\omega)d\nu$ maximizes profit relative to p on the total production set.

A theorem which establishes the existence of competitive allocations in the case of an exchange economy has been given by R. J. AUMANN [2].

We shall now exhibit the connection between the core and the set of competitive allocations. The relevance of this connection to economic theory lies in the fact that the allocations in the core are determined only through the blocking process without a price system associated with commodities while the competitive allocations are obtained by the adaptation of all economic agents to a suitably chosen price system. The first concept is a refinement of the well-known Pareto optimality and assumes complete information and free coalition formation while the second concept assumes that every economic agent adapts himself to (i.e., does not try or has not the power to influence) the prevailing price system.

<u>THEOREM 1: Every competitive allocation in an economy with a measure space of economic agents belongs to the core.</u>

<u>PROOF:</u> Let x be a competitive allocation with respect to the price system p. Assume that the allocation x is blocked, i.e., there exist a coalition $E \in \mathcal{A}$ and an allocation z such that

(i) $x_a <_a z_a$ for almost every $a \in E$,

(ii) $\nu(E) > 0$ and $\int_E z d\nu \in \int_E(\omega+Y)d\nu$.

By (C.1) of the definition of a competitive allocation (i) implies $(p|\omega_a) + \sup(p|Y_a) < (p|z_a)$ for almost every $a \in E$. Hence

$$\int_E \sup(p|Y)d\nu < \int_E(p|z-\omega)d\nu = (p|\int_E(z-\omega)d\nu).$$

Since

$$\int_E \sup(p|Y)d\nu = \sup(p|\int_E Yd\nu)$$

we have

$$\sup(p|\int_E Yd\nu) < (p|\int_E(z-\omega)d\nu).$$

But this is a contradiction since by (ii) it follows that $\int_E(z-\omega)d\nu \in \int_E Yd\nu$.

An attainable allocation x is said to be *quasi-competitive*, if there is a price system $p \in S$, $p \neq 0$, such that

(Q.1) for almost every $a \in A$, x_a is a maximal element with respect to $<_a$ in the budget set $\gamma_a(p)$ and/or $\inf(p|X_a) = (p|\omega_a) + \sup(p|Y_a)$.

(C.2) $(p|\int_A (x-\omega)d\nu) = \max\, (p|\int_A Y d\nu)$.

THEOREM 2: Let $\mathcal{E}$ *be an economy with a measure space of economic agents such that the following assumptions hold:*

(IV) *Pure competition.* The measure space $(A, \mathfrak{a}, \nu)$ is atomless, i.e., for every coalition $E \in \mathfrak{a}$ with $\nu(E) > 0$ there is a coalition $M \in \mathfrak{a}$ with $M \subset E$ such that $0 < \nu(M) < \nu(E)$. (This assumption implies that no individual economic agent can, by his own decision, influence the outcome of the collective activity.)

(V) *Local nonsatiation.* Almost everywhere in A, if $z \in X_a$ is not a maximal element of X_a then for every neighborhood $\mathcal{U}(z)$ of z there is a vector $x' \in \mathcal{U}(z) \cap X_a$ with $z <_a x'$.

(VI) *Continuous preference relations.* Almost everywhere in A, for every $z \in X_a$ the set $\{x \in X_a | z \quad_a x\}$ is open in X_a.

(VII) *Convex consumption sets.* Almost everywhere in A, the consumption set X_a is a convex subset in S.

Every allocation in the core for which almost every economic agent is not satiated is quasi-competitive.

PROOF: Let the allocation x belong to the core and assume that x_a is not a satiation consumption for almost every $a \in A$. The first step in the proof consists in determining a price vector p, then we show that x is an expenditure minimizing allocation with respect to p and finally we prove that the allocation x is quasi-competitive with respect to p.

Consider the correspondence Ψ resp. ϕ of A into S defined by

$$\Psi a = \{z \in X_a | x_a <_a z\}$$

$$\text{resp.} \quad \phi_a = \Psi_a - \omega_a.$$

We denote by $\mathcal{L}_\phi$ the set of all ν-integrable functions f of A into S such that $f_a \in \phi_a$ for almost every $a \in A$ (analogously, we define $\mathcal{L}_\Psi$ and $\mathcal{L}_Y$).

Using assumptions (II) and (V) and the Measurable Selection Theorem it follows easily

(1) $$\mathcal{L}_\phi \neq \emptyset.$$

Consider now the set $Z \subset S$ defined by

$$Z = \{\int_E (f-g)d\nu | E \in \mathcal{A} \text{ with } \nu(E) > 0, f \in \mathcal{L}_\phi, g \in \mathcal{L}_Y\}.$$

We shall now show that

(2) $$0 \notin Z,$$

(3) $$Z \text{ is convex.}$$

To (2): Assume $0 \in Z$, i.e., there exist an $E \in \mathcal{A}$ with $\nu(E) > 0$ and a function $f \in \mathcal{L}_\phi$ such that $\int_E f d\nu \in \int_E Y d\nu$. But then the coalition E would block the allocation x since the allocation $h = f + \omega$ has the properties: $x_a <_a h_a$ for almost every a in E and $\int_E h d\nu \in \int_E (\omega + Y)d\nu$.

To (3):[6] Denote $\mathcal{L} = \mathcal{L}_\phi - \mathcal{L}_Y$. Let $z_i = \int_{E_i} h_i d\nu$ with $\nu(E_i) > 0$ and $h_i \in \mathcal{L}$, $i = 1,2$. We show that for each real number λ with $0 < \lambda < 1$ there exist a set $E \in \mathcal{A}$ with $\nu(E) > 0$ and a function $h \in \mathcal{L}$ such that

$$z = \lambda z_1 + (1-\lambda)z_2 = \int_E h d\nu.$$

Denote by $h_i \cdot \nu$ the S-valued measure defined by

$$(h_i \cdot \nu)(E) = \int_E h_i d\nu, \qquad i = 1,2$$

and $(h_1, h_2) \cdot \nu$ the $S \times S$-valued measure defined by

$$((h_1, h_2) \cdot \nu)(E) = (\int_E h_1 d\nu, \int_E h_2 d\nu).$$

[6] The proof of assertion (3) is analogous to Lemma A in K. VIND [9].

Applying the Theorem of Liapunov to the atomless measure spaces $(E_1 \setminus E_2, \mathfrak{A} \cap (E_1 \setminus E_2), h_1 \cdot \nu)$ resp. $(E_2 \setminus E_1, \mathfrak{A} \cap (E_2 \setminus E_1), h_2 \cdot \nu)$ resp. $(E_1 \cap E_2, \mathfrak{A} \cap E_1 \cap E_2, (h_1, h_2) \cdot \nu)$ there follows the existence of measurable sets $B_1 \subset E_1 \setminus E_2$ resp. $B_2 \subset E_2 \setminus E_1$ resp. $B_3 \subset E_1 \cap E_2$ such that

$$\int_{B_1} h_1 d\nu = \lambda \int_{E_1 \setminus E_2} h_1 d\nu \quad \text{resp.} \quad \int_{B_2} h_2 d\nu = (1-\lambda) \int_{E_2 \setminus E_1} h_2 d\nu$$

$$\text{resp.} \quad \left(\int_{B_3} h_1 d\nu, \int_{B_3} h_2 d\nu\right) = \lambda \left(\int_{E_1 \cap E_2} h_1 d\nu, \int_{E_1 \cap E_2} h_2 d\nu\right).$$

Define now

$$h(a) = \begin{cases} h_1(a) & \text{for } a \in B_1 \cup B_3 \\ h_2(a) & \text{for } a \in A \setminus (B_1 \cup B_3) \end{cases}$$

and

$$E = B_1 \cup B_2 \cup (E_1 \cap E_2).$$

One verifies easily

$$\int_E h d\nu = \lambda \int_{E_1} h_1 d\nu + (1-\lambda) \int_{E_2} h_2 d\nu.$$

Since $h \in \mathcal{L}$ and $E \in \mathfrak{A}$ with $\nu(E) > 0$ we have proven the convexity of the set Z.

By Minkowski's Separation Theorem there exists a vector $p \in S$, $p \neq 0$, such that

(4) $\qquad (p|\int_E g d\nu) \leqq (p|\int_E f d\nu)$ for every $E \in \mathfrak{A}$, $g \in \mathcal{L}_Y$ and $f \in \mathcal{L}_\phi$

or

(4') $\qquad \sup (p|\int_E Y d\nu) + (p|\int_E \omega d\nu) \leqq (p|\int_E h d\nu)$ for every $E \in \mathfrak{A}$ and $h \in \mathcal{L}_\psi$,

i.e., no coalition can improve almost every member compared with x_a spending less than the total wealth of the coalition.

Next we show that (4) implies

(5) $\qquad$ for almost every $a \in A$,

$(p|\omega_a) + \sup(p|Y_a) \leqq (p|z)$ for every $z \succ_a x_a$, i.e., the value of any consumption plan preferred by agent a to x_a is greater or equal to his wealth.

It is not difficult to show that the set

$$\{a \in A \mid (p|\omega_a) + \sup\,(p|Y_a) \leqq (p|\phi_a)\}$$

belongs to $\mathcal{A}_\nu$. Consequently, if (5) does not hold, there exists a set $B \in \mathcal{A}$ with $\nu(B) > 0$ such that for every $a \in B$ there is a point $z \in \psi_a$ with $(p|z) < (p|\omega_a) + \sup(p|Y_a)$. Using the Measurable Selection Theorem one can easily derive a contradiction to (4).

The allocation x is an expenditure minimizing allocation if, in addition to (5), we show that the consumption plan x_a belongs to the budget set $\gamma_a(p)$ and that profit is maximized.

More specifically, we show

(6) $$\text{for almost every } a \in A, \quad (p|x_a) = (p|\omega_a) + \sup(p|Y_a).$$

Since by assumption (V) almost everywhere in A the vector x_a belongs to the closure of ψ_a it follows by (5): $(p|\omega_a) + \sup(p|Y_a) \leqq (p|x_a)$.

Assume $(p|\omega_a) + \sup(p|Y_a) < (p|x_a)$ for every $a \in M$ with $\nu(M) > 0$. Then $\int_A (p|\omega)d\nu + \int_A \sup(p|Y)d\nu < \int_A (p|x)d\nu$. Hence

$$\sup(p|\textstyle\int_A Y d\nu) < (p|\int_A (x-\omega)d\nu).$$

Since x is an attainable allocation, i.e., $\int_A(x-\omega)d\nu \in \int_A Y d\nu$, we have a contradiction. This proves (6).

Part (C.2) of the theorem follows immediately from statement (6), since

$$\sup(p|\textstyle\int_A Y d\nu) = \int_A \sup(p|Y)\, d\nu = (p|\int_A (x-\omega)d\nu).$$

To complete the proof it remains to show that x is quasi-competitive, i.e., we have to prove that in the case $\inf\,(p|X_a) < (p|\omega_a) + \sup(p|Y_a)$, the consumption plan x_a is a maximal element in the budget set. By property (5) we have

$$z \in X_a \text{ and } (p|z) < (p|\omega_a) + \sup(p|Y_a) \text{ implies } z \succ_a x_a.$$

Let z be any point in the budget set $\gamma_a(p)$. Then z can be obtained as a limit of points z_n, $n = 1, \ldots,$ with $(p|z_n) < (p|\omega_a) + \sup(p|Y_a)$ since X_a is assumed to be convex. Hence the continuity assumption (VI) implies $z \succ_a x_a$ which proves that x_a is a maximal element in the budget set.

REMARK: The use of measure theory in mathematical economics in order to describe an economy where no individual economic agent can influence the outcome of the collection activity is due to R. J. AUMANN [1]. AUMANN proved the above theorem in the case $X_a = R_+^\ell$ (positive orthant), $Y_a = \{0\}$ (pure exchange) and monotone preferences $\prec_a$, i.e., every commodity is desired. In this situation the allocations in the core are actually competitive (not just quasi-competitive). An alternative proof for Aumann's result has been given by K. VIND [9].

REFERENCES

[1] AUMANN, R. J.,
Markets with a Continuum of Traders. Econometrica, 32, 39-50 (1964).

[2] AUMANN, R. J.,
Existence of Competitive Equilibria in Markets with a Continuum of Traders. Econometrica, 34, 1-17 (1966).

[3] AUMANN, R. J.,
Measurable Utility and Measurable Choice Theorem. (To appear in: Proceedings of the Colloque International du Centre National de la Recherche Scientifique "La Decision", Aix-en-Provence, 3-7 Juillet 1967).

[4] DEBREU, G.,
Theory of Value. New York, Wiley (1959).

[5] HILDENBRAND, W.,
The Core of an Economy with a Measure Space of Economic Agents. (To appear in The Review of Economic Studies, Oct. 1968).

[6] LIAPUNOV, A.,
Sur les Fonctions-vecteurs Completement Additives. Bull. Acad. Sci. URSS, Ser. Math., 4, 465-478 (1940).

[7] VON NEUMANN, J.,
On Rings of Operators, Reduction Theory. Ann. Math., 50, 401-485 (1949).

[8] SION, M.,
On Uniformization of Sets in Topological Spaces. Trans. Am. Math. Soc., 96, 237-246 (1960).

[9] VIND, K.,
Edgeworth-Allocations in an Exchange Economy with Many Traders. International Economic Review, 5, 165-177 (1964).

GEOMETRIC THEORY OF LINEAR CONTROLLED SYSTEMS*

E. B. Lee

Introduction

Geometric techniques are widely used in the study of optimal control problems. In the proof of the maximum principle [1] the essential arguments involved geometric ideas such as the cone of attainability and the existence of a separating hyperplane in a certain space. The generation of this cone depended on linearization and the proof of the maximum principle then followed from the linear theory by showing that the linearized cone gave a good approximation (essentially a fixed point result of the type given by BROUWER was all that was needed). In this paper geometric ideas will be exploited to obtain results from which necessary and sufficient conditions for optimal control follow for linear models and to establish an existence theorem for optimal control. The necessary condition, the maximum principle, can be established for certain nonlinear models using the geometric results and a fixed point argument.

The mathematical control model considered is the linear functional-differential equation

$$\mathscr{L}) \qquad \dot{x}(t) = \int_{-\tau}^{0} \{[d_s\, A(t,s)]x(t+s) + [d_s\, B(t,s)]u(t+s)\}$$

with continuous initial function $x(t) = \phi(t)$ on $[t_0-\tau, t_0]$. Here

(i) $x(t)$ is an $n \times 1$ state vector;

(ii) $u(t) \in \Omega$ is an $m \times 1$ vector control function;

(iii) $\Omega \subset R^m$ is a compact convex restraint set;

(iv) $\tau > 0$ is a constant;

---

* Research sponsored by Air Force Office of Scientific Research, Office of Aerospace Research, United States Air Force, Grant No. AF-AFOSR-571-66.

(v) $A(t,s)$ defined for $t \geq t_0$, $-\infty < s < \infty$ is an $n \times n$ matrix continuous in t uniformly with respect to s, $-\tau \leq s \leq 0$. Each element of $A(t,s)$ is of bounded variation with respect to s, $-\tau \leq s \leq 0$;

(vi) $B(t,s)$ is an $n \times m$ matrix having the properties of $A(t,s)$ given in v ;

(vii) R^n is the n dimensional real number space.

The integral is in the sense of Lebesgue-Stieltjes. Note that equation $\mathscr{L}$ can be written in the form of a delay-differential equation*

$$\mathscr{D}) \qquad \dot{x}(t) = \sum_{i=0}^{\ell} A_i(t)x(t-h_i) + \int_{-\tau}^{0} K(s,t)x(t+s)ds$$

$$+ \sum_{j=0}^{\gamma} B_i(t)u(t-k_i) + \int_{-\tau}^{0} L(s,t)u(t+s)ds$$

if, in the decomposition of $A(s,t)$ (and of $B(s,t)$) into a continuous function of bounded variation and a Saltus function, the Saltus function depends only on the difference $t-s$ [2]. The linear control system without delays, which has been studied so extensively in the literature, is obtained by assuming all terms $A_1(t)\ldots A_\ell(t)$, $K(s,t)$, $B_1(t)\ldots B_\gamma(t)$ and $L(s,t)$ are zero. Systems with no delay in the control variables are obtained from $\mathscr{D}$ by assuming just $B_1(t)\ldots B_\gamma(t)$ and $L(s,t)$ are zero. The linear control problems without delays in the control variables were considered in [3].

Let $Y(s,t)$ be the $n \times n$ matrix solution [3] of

$$Y(s,t) + \int_{s}^{s+\Delta} Y(\sigma,t)A(\sigma,s-\sigma)d\sigma = I$$

where

$$\Delta = t-s \quad \text{if} \quad t-\tau \leq s \leq t$$

$$\tau \quad \text{if} \quad t_0 \leq s \leq t-\tau$$

and I is the $n \times n$ identity matrix. Further, let $X(t,\phi)$ be the solution [3] of the homogeneous equation

$$\dot{x}(t) = \int_{-\tau}^{0} d_s A(t,s)x(t+s)$$

with $x(t) = \phi(t)$ on $t_0 - \tau \leq t \leq t_0$. Then given a measurable function

---

* $0 = h_0 \leq h_1 < h_2 \ldots h_\ell \leq \tau$, $0 = k_0 \leq k_1 < \ldots k_\gamma \leq \tau$

$u(s) \subset \Omega$ on $t_0 - \tau \leq s \leq t$ the corresponding response $x(t)$ of $\mathscr{L}$ can be obtained from

$$(\mathscr{L}) \qquad x(t) = X(t,\phi) + \int_{t_0}^{t} Y(s,t) \{\int_{-\tau}^{0} d_\sigma B(s,\sigma)u(s+\sigma)\}ds,$$

for $t \geq t_0$.

To facilitate the discussion and to obtain a pointwise maximum principle it is assumed for the remainder of this paper that only one delayed control term occurs in the model. Namely, consider the system

$$(\mathscr{D}_1) \qquad \dot{x}(t) = \int_{-\tau}^{0} d_s A(t,s)x(t+s) + B_0(t)u(t) + B_1(t)u(t-h)$$

with $\tau \geq h \geq 0$.

All arguments of course apply to the case when there are a finite number of delayed control terms and even to the case of an integral term involving the controller as in equation $\mathscr{D}$. This special form is needed so that the order of integration in equation $\mathscr{L}$ can be easily interchanged and the pointwise maximum principle obtained. In a future paper the general system $\mathscr{L}$ will be treated. The specialization of equation $\mathscr{L}$ to the system $\mathscr{D}_1$ gives

$$(\mathscr{L}_1) \qquad x(t) = X(t,\phi) + \int_{t_0}^{t} Y(s,t) \{B_0(s)u(s) + B_1(s)u(s-h)\}ds$$

$$= X(t,\phi) + \int_{t_0-h}^{t} H(s,t)u(s)ds$$

where

$$H(s,t) = Y(s+h,t)B_1(s+h) \quad \text{on} \quad t_0-h \leq s \leq t_0$$

$$= Y(s,t)B_0(s) + Y(s+h,t)B_1(s+h) \quad \text{on} \quad t_0 \leq s \leq t-h$$

$$= Y(s,t)B_0(s) \quad \text{on} \quad t-h \leq s \leq t .$$

It is assumed that the action of the controllers realized by the response of the model $\mathscr{D}_1$ are evaluated by a cost functional of control

$$C(u) = g(x(t_1)) + \int_{t_0}^{t_1} \{f^0(x(t),t) + h^0(u(t),t)\}dt .$$

In addition it is assumed that there is a side constraint of the form

$$\int_{t_0}^{t_1} \{f^{n+1}(x(t),t) + h^{n+1}(u(t),t)\}dt \leq \beta ,$$

and that there is a possible requirement that the response end point $x(t_1)$ lie in a prescribed closed target set $G \subset R^n$ or that $x(t) = \psi(t)$, a given function, for $t_1 - \tau \leq t \leq t_1$. If the cost functional includes the initial interval $[t_0-h,t_0]$ a slight refinement of the results obtained is needed.

$f^0(x,t)$, and $f^{n+1}(x,t)$ are C^1 real-non-negative convex functions of $x \in R^n$ for all t . $h^0(u,t)$, and $h^{n+1}(u,t)$ are real-non-negative convex continuous functions of $u \in R^m$ for all t . $g(x)$ is a C^1 convex function of $x \in R^n$ and $\beta > 0$ is a constant. These conditions can be relaxed in certain of the results discussed below, for example, the necessary condition for optimal control does not use the convexity hypothesis. Additional side constraints of the above inequality or equality type are readily handled by the techniques described here.

The problem of optimal control is to select measurable $u(t) \subset \Omega$ on $[t_0-h,t_1]$ (or just on $[t_0,t_1]$ if the initial control function is prescribed on $[t_0-h,t_0]$) to steer the corresponding response $x(t)$ of $\mathcal{D}_1$ from the continuous initial function $\phi(t)$ to the closed target set G at time t_1, (or to the given final function $\psi(t)$ on $[t_1-\tau,t_1]$) while satisfying the given side constraint and such that the cost functional of control is minimized.

In some problems the initial function $\phi(t)$ may also be selected in the minimization. This can be handled in the formulation which follows if $\phi(t) \subset M \subset R^n$ on $[\tau-t_0, t_0]$ is measurable with M compact by expressing the solution of the homogeneous equation $X(t,\phi)$ in an integral form involving $\phi(t)$, [see 4, page 360].

Let $\tilde{x}(t) = (x^0(t), x(t), x^{n+1}(t)) \in R^{n+2}$ be an $n+2$ vector and consider the system model

$$\tilde{\mathcal{D}}_1) \qquad \begin{aligned} \dot{x}^0(t) &= f^0(x(t),t) + h^0(u(t),t) \\ \dot{x}(t) &= \int_{-\tau}^{0} d_s\, A(t,s)x(t+s) + B_0(t)u(t) + B_1(t)u(t-h) \\ \dot{x}^{n+1}(t) &= f^{n+1}(x(t),t) + h^{n+1}(u(t),t) \end{aligned}$$

with initial data $\tilde{x}(t) = \tilde{\phi}(t) = (0,\phi(t),0)$ on $[t_0-\tau,t_0]$. Further, let

$$\tilde{G} = \{(x^0,x,x^{n+1}) | -\infty < x^0 < \infty,\ x \in G,\ x^{n+1} \leq \beta\} .$$

The optimal control problem can now be stated as: Select a measurable controller $u(t) \subset \Omega$ on $[t_0-h,\ t_1]$ such that it steers the corresponding response $\tilde{x}(t)$ of $\tilde{\mathcal{D}}_1$ from $\tilde{\phi}(t)$ on $[t_0-\tau,t_0]$ to the closed target set $\tilde{G}$ at time t_1 and minimizes the functional $C(u) = g(x(t_1)) + x^0(t_1)$.

Geometric Theory

The set of attainability $\tilde{K}(t_1) \subset R^{n+2}$ is the collection of end points $\tilde{x}(t_1) \in R^{n+2}$ of responses $\tilde{x}(t)$ of $\tilde{\mathcal{D}}_1$ with continuous initial function $\tilde{\phi}(t)$ on $[t_0-\tau,\ t_0]$ corresponding to all measurable controllers $u(s) \subset \Omega$ on $[t_0-h,\ t_1]$. The saturation set $\tilde{K}_s(t_1)$ of $\tilde{K}(t_1)$ is the collection of all points $\tilde{y} = (y^0,y,y^{n+1}) \in R^{n+2}$ for which there is a point $(x^0,x,x^{n+1}) \in \tilde{K}(t_1)$ with

$$x^0 \leq y^0 \ ,\ x = y,\ x^{n+1} \leq y^{n+1} .$$

Apparently $\tilde{K}(t_1) \subset \tilde{K}_s(t_1) \subset \{(x^0,x,x^{n+1}) | x^0 \geq 0,\ x \in R^n\ x^{n+1} \geq 0\}$. Moreover, $\tilde{K}(t_1)$ lies in some sphere $\tilde{S}(r) \subset R^{n+2}$ of finite radius r centered at the origin.

In [5] it is established that $\tilde{K}_s(t_1) \subset R^{n+2}$ is closed and convex for a linear ordinary differential equation model with cost functional and constraint as above. No modification of that proof is needed to establish the closedness and convexity of $\tilde{K}_s(t_1)$ for the above linear functional differential equation model.

Using these properties of $\tilde{K}_s(t_1)$ it is straightforward [5] to prove:

THEOREM 1 Consider the system $\tilde{\mathcal{D}}_1$ with saturation set $\tilde{K}_s(t_1)$, closed target set $\tilde{G}$, and cost functional $C(u) = g(x(t_1)) + x^0(t_1)$. If there exists a measurable controller $u(t) \subset \Omega$ on $[t_0-h,\ t_1]$ such that the corresponding response $\tilde{x}(t)$ of $\tilde{\mathcal{D}}_1$ with $\tilde{x}(t) = \tilde{\phi}(t)$ on $[t_0-\tau_1,t_0]$ has an end point $\tilde{x}(t_1) \in \tilde{G}$ then there exists an optimum controller. Moreover, the end point $\tilde{x}^*(t_1)$ of the response corresponding to the optimum controller $u^*(t)$ on $[t_0-h,\ t_1]$ is a point of the boundary of the saturation set $\tilde{K}_s(t_1)$.

Since the optimum points of the set of attainability or its saturation are necessarily boundary points it is important to characterize such boundary points in terms of the controllers which have such response end points.

An admissible controller $u^*(t) \subset \Omega$ on $[t_0-h, t_1]$ is a <u>maximal controller on</u> $[t_0-h, t_1]$ if there is a nontrivial solution $\tilde{\eta}(t) = (\eta_0(t), \eta(t), \eta_{n+1}(t))$, $n+2$ row vector, of

$$\tilde{a}) \quad \begin{aligned} &\eta_0(t) \equiv \eta_0 = \text{constant} \leq 0 \\ &\eta_{n+1}(t) \equiv \eta_{n+1} = \text{constant} \leq 0 \\ &\eta(t) + \int_t^{t+\Delta} \eta(s)\, A(s,t-s)ds \\ &\qquad + \int_{t_1}^{t} \{\eta_0 \frac{\partial f^0}{\partial x}(x^*(s),s) + \eta_{n+1} \frac{\partial f^{n+1}}{\partial x}(x^*(s),s)\}ds \\ &= \text{constant} \end{aligned}$$

$$\Delta = \begin{cases} t_1-t & \text{if } t_1-\tau \leq t \leq t_1, \\ \tau & \text{if } t_0 \leq t \leq t_1-\tau \end{cases}$$

such that

$$\eta(t)B_0(t)u^*(t) + \eta_0 h^0(u^*(t),t) + \eta_{n+1} h^{n+1}(u^*(t),t)$$

$$= \max_{u\in\Omega} \{\eta(t)B_0(t)u + \eta_0 h^0(u,t) + \eta_{n+1} h^{n+1}(u,t)\}$$

almost everywhere on $[t_1-h, t_1]$

$$\eta(t)B_0(t)u^*(t) + \eta(t+h)B_1(t+h)u^*(t) + \eta_0 h^0(u^*(t),t)+\eta_{n+1}h^{n+1}(u^*(t),t)$$

$$= \max_{u\in\Omega} \{\eta(t)B_0(t)u + \eta(t+h)B_1(t+h)u + \eta_0 h^0(u,t) + \eta_{n+1}h^{n+1}(u,t)\}$$

almost everywhere on $[t_0, t_1-h]$

$$\eta(t+h)B_1(t+h)u^*(t) = \max_{u\in\Omega} \{\eta(t+h)B_1(t+h)u\}$$

almost everywhere on $[t_0-h, t_0]$.

Here $x^*(t)$ is the response of $\mathcal{D}_1$ corresponding to the admissible controller $u^*(s)$ on $[t_0-h, t]$. A controller $u(s)$ is admissible on $[t_0-h, t]$ if it is measurable thereon with $u(s) \in \Omega$, for $t_0-h \leq s \leq t$.

THEOREM 2 If an admissible controller $u(t)$ is a maximal controller on $[t_0-h, t_1]$ the corresponding response end point $\tilde{x}(t_1) \in \partial \tilde{K}_s(t_1)$. Conversely, if $\tilde{x}_1 \in \tilde{K}$ is a point of $\partial \tilde{K}_s(t_1)$ then the controller with response end point $\tilde{x}(t_1) = \tilde{x}_1$ is a maximal controller.

The technique of proof of this result is now standard [5] and involves observing for the above system $\tilde{\mathcal{D}}_1$ that if $\tilde{\eta}(t)$ is a solution of $\tilde{\alpha}$ corresponding to a maximal controller $u(s)$ on $[t_0-h, t]$ and system response $\tilde{x}(t)$ then

$$\tilde{\eta}(t_1)\tilde{x}(t_1) - \tilde{\eta}(t_0)\tilde{x}(t_0) = \int_{t_0}^{t_1} \eta_0[f^0(x(t),t) - \frac{\partial f^0}{\partial x}(x(t),t)x(t)]dt$$

$$+ \int_{t_0}^{t_1} \eta_{n+1}[f^{n+1}(x(t),t) - \frac{\partial f^{n+1}}{\partial x}(x(t),t)x(t)]dt$$

$$+ \int_{t_0}^{t_1} \{\eta_0 h^0(u(t),t) + \eta_{n+1} h^{n+1}(u(t),t) - \eta(t)B_0(t)u(t)\}dt$$

$$+ \int_{t_0-h}^{t_1-h} \eta(t+h)B_1(t+h)u(t)dt + \int_{t_0-\tau}^{t_0} d_s\{\int_{t_0}^{s+\sigma} \eta(t)A(t,s-t)dt\}x(s)$$

$$\sigma = t_1 - s \quad \text{if} \quad t_1 - \tau \leq s \leq t_0$$

$$= \tau \quad \text{if} \quad t_0 - \tau \leq s \leq t_1 - \tau$$

(Note the first case does not occur if $t_0+\tau \leq t_1$)

and if $\tilde{x}_s(t)$ is any other response $\tilde{\eta}(t_1)[\tilde{x}(t_1) - \tilde{x}_s(t_1)] \geq 0$.

Using the above properties of the set of attainability and its saturation necessary and sufficient conditions for optimum controllers (with tranversality conditions) are easily established using the procedures outlined in [5] and [6].

REFERENCES

[1] PONTRYAGIN, L. S.; BOLTYANSKII, V. G.; GAMKRELIDZE, R. V.; MISCHENKO, E. F., The Mathematical Theory of Optimal Processes. Interscience, New York (1962).

[2] OGUZTÖRELI, M. N., Time-Lag Control Systems. Academic Press, New York (1966).

[3] CHYUNG, D.; LEE, E. B., On certain extremal problems involving linear functional differential equation models. In Proceeding Mathematical Theory of Control, Editors L. NEUSTADT and A. V. BALAKRISHNAN, Academic Press, New York (1967).

[4] HALANAY, A., Differential equations - Stability, Oscillations, Time Lags. Academic Press, New York (1965).

[5] LEE, E. B., Linear optimal control problems with isoperimetric constraints. IEEE trans. on Auto Control, Vol. AC-12, No. 1, 87-90 (1967).

[6] LEE, E. B.; MARKUS, L., Foundations of Optimal Control Theory. John Wiley and Sons Inc., New York (1967).

STABILITY OF SETS WITH RESPECT TO ABSTRACT PROCESSES

Jozef Nagy

Introduction

In the study of ordinary differential equations we are usually concerned with situations in which most of the following conditions (related to the initial-value problem) are satisfied:

(i) local existence of solutions;

(ii) indefinite prolongability of solutions;

(iii) unicity of solutions;

(iv) autonomness.

In generalizing the concept of an ordinary differential equation, the axiomatic approach to differential equation theory has been employed by many authors. Thus, in analogy with equations satisfying (i) to (iv), MARKOV defined the dynamical systems in metric spaces [17]. A very profound and exhaustive description of properties of these dynamical systems in metric and uniform spaces is presented in the well-known book [6] by GOTTSCHALK and HEDLUND; however, systems with properties (i), (iii) and (iv) are treated in URA's paper [23] and HÁJEK's book [7]. There are global flows [7], corresponding to equations satisfying (i), (ii) and (iii), and local flows investigated in [8], [14], and flows studied by KURZWEIL [12], satisfying (i) and (iii). To equations, which satisfy (i) and (ii), there correspond the generalized systems of ZUBOV [25], the generalized control system of ROXIN [18], dynamical polysystems of BUSHAW and HALKIN [11], and finally, to equations satisfying (i) only, there correspond the local generalized control systems of ROXIN [19].

On the 2nd EQUADIFF Symposium, held 1966 in Bratislava, HÁJEK suggested to eliminate all the properties (i) to (iv) and defined an abstract process on an abstract set [9], [10]. This is, to my best knowledge, the last step in the generalization of differential equation concept. The definition and several elementary properties of the process are described in the first part of this paper.

A great deal of papers on dynamical systems and flows is concerned with stability properties of these objects. URA's results on prolongations in dynamical systems were used by himself [23], [24], AUSLANDER, BHATIA, SEIBERT, SZEGÖ [1], [2], [3], [4], [5], [22] for studying certain boundedness and stability properties of dynamical systems. Liapunov's functions and generalized Liapunov's functions were employed by BHATIA [4], HÁJEK [7], AUSLANDER and SEIBERT [2] and others in the dynamical systems theory, by ZUBOV [25] in the generalized systems theory, by NAGY [14], [15] in local flows, and by ROXIN [20], [21] in the control systems theory. Very profound results on the stability of integral manifolds of local flows in metric spaces are contained in KURZWEIL's papers [12], [13]. In this paper LIAPUNOV's functions are used for discussing certain strong, weak and asymptotic stability properties of sets with respect to an abstract process on an arbitrary set (along with corresponding uniform modifications).

In what follows, R^n, R^+ and R^o denote the Euclidean n-space, the set of all positive reals and the set of all non-negative reals, respectively. The cartesian product of sets X and Y is denoted by $X \times Y$.

Now, let us recall several elementary facts concerning the concept of a relation between two sets. A relation between sets X and Y is a subset of $X \times Y$; a relation on X is a relation between X and X. If r is a relation between X and Y and $x \in X$ and $y \in Y$ are in the relation r, we prefer to write xry instead of $(x,y) \in r$. The relation inverse to r is denoted by r^{-1}; thus xry iff $yr^{-1}x$. The identity relation 1 on X is defined by x1y iff $x=y \in X$. For some pair of relations r, r′ the composition $r \circ r'$ is defined by $x\, r \circ r' y$ iff xru and ur′y for some u. A relation r between Y and X will be termed a partial map iff $r \circ r^{-1} \subset 1$ (and this will be abbreviated by: partial map r: $X \to Y$). It may be useful to emphasise that the composition operation $\circ$ is even for partial maps; that the value of a partial map f at an element x is written as fx, with round parantheses used primarily for complying with the usual conventions on precedence. Finally, let us recall the fact that with each map $f: X \to Y$ we can associate, in a one-to-one manner, the relation between Y and X denoted again by f and defined by yfx iff $y = fx$. The natural composition of maps corresponds then to the natural composition of the corresponding relations.

1. Abstract processes

1.1 NOTATION. Assume given a set P, and also a subset $R \subset R^1$. The set R is ordered by the natural order relation inherited from R^1, and in particular one may speak of intervals in R (i.e. the order-convex subsets of R^1; this includes the void set and singletons, i.e. degenerate intervals). Consider any relation p on P×R, i.e. between P×R and P×R with the property that

(1.1.1) $$(y,\beta)p(\omega,\alpha) \text{ implies } \beta \geq \alpha .$$

Then p defines a system $\{{}_\beta p_\alpha : \beta \geq \alpha\}$ of relations ${}_\beta p_\alpha$ on P, in the following manner:

(1.1.2) $$y\,{}_\beta p_\alpha x \text{ iff } (y,\beta)p(x,\alpha)$$

These ${}_\beta p_\alpha$ will be called the individual relations of p. It is obvious that, conversely, any system $\{{}_\beta p_\alpha : \beta \geq \alpha\}$ of relations on P defines, by (1.1.2), a relation p of the type described above. This notation will be used to set up the following definition.

1.2 DEFINITION. p is a <u>process</u> <u>on</u> P <u>over</u> R iff P is a set, $R \subset R^1$, p is a relation on P×R with (1.1.1) and the following two conditions are satisfied:

(i) $${}_\alpha p_\alpha \subset 1 \quad \text{for all} \quad \alpha \in R;$$

(ii) $${}_\gamma p_\beta \circ {}_\beta p_\alpha = {}_\gamma p_\alpha \quad \text{for all} \quad \gamma \geq \beta \geq \alpha \quad \text{in R.}$$

1.3 REMARK. 1.2(i) may be termed the initial value property, and 1.2(ii) the compositivity property. In what follows, given a process p on P over R, the symbols $D, E, {}_\theta p_\alpha x$ will be used to denote the sets

$$D = \text{domain } p = \{(x,\alpha) \in P\times R: x\,{}_\alpha p_\alpha x\};$$

$$E = \{(\theta,x,\alpha) \in R\times P\times R: y\,{}_\theta p_\alpha x \quad \text{for some} \quad y \in P\};$$

$$_{\theta}p_{\alpha}x = \{y \in P:\ y_{\theta}p_{\alpha}x \ \text{ for a given } (\theta,x,\alpha) \in E\}.$$

Clearly, each process p is a partial order on the corresponding D; it is this partial order which is meant when one says that a function e. g. increases along p. More precisely, a partial map $V:P\times R \to R$ is termed increasing along a process p in P iff $V|D$ is an increasing map between D (endowed with p as a partial order) and R. Similarly for non-increasing maps, etc.

To exhibit the basic interpretation of Definition 1.2, we shall describe the following examples.

1.4 EXAMPLES

a) Consider a differential equation

$$\frac{dx}{d\theta} = f(x,\theta) \tag{1.4.1}$$

in euclidean n-space R^n, with continuous $f: D \to R^n$ and D an open set in R^{n+1}. The classical solutions of (1.4.1) are partial maps $s: R^1 \to R^n$ such that domain s is an interval in R^1, either degenerate or with

$$\frac{d}{d\theta}\ s\theta = f(s\theta,\theta) \quad \text{for all} \quad \theta \in \text{domain } s$$

With the differential equation (1.4.1) one associates the process p, in R^n, defined by setting $y_{\beta}p_{\alpha}x$ (for x,y in R^n, $\alpha \le \beta$ in R^1) iff there is a classical solution of (1.4.1) assuming the values x and y at α and β, respectively. Processes obtained in this manner, with the exhibited assumptions on f, are termed differential.

b) As a less immediate interpretation, consider a difference-differential equation with constant time lag

$$\frac{dx}{d\theta} = f(x\ (\theta-\tau),\ x(\theta),\ \theta), \tag{1.4.2}$$

given continuous $f : R^3 \to R^1$ and $\tau > 0$. For definiteness, the solutions of (1.4.2) are continuous maps $s : [\alpha-\tau,\beta] \to R^1$ for given $-\infty < \alpha \le \beta < +\infty$ such that

$$\frac{d}{d\theta}\ s(\theta) = f(s(\theta-\tau),\ s(\theta),\theta) \quad \text{for} \quad \alpha \le \theta < \beta$$

(with obvious modifications for the case of non-closed domains). It will be convenient to write x_λ for the λ-translate of a partial map $x : R^1 \to R^1$, so that $x_\lambda(\theta) = x(\theta + \lambda)$ whenever defined. The initial value problem for (1.4.2) is to find, to given $\alpha \in R^1$ and continuous $y: \langle -\tau, o \rangle \to R^1$, a solution s of (1.4.2) as above, and satisfying $y \subset s_\alpha$, i.e. such that $s(\theta) = y(\theta - \alpha)$ for $\alpha - \tau \le \theta \le \alpha$. This situation may be usefully described by a process p in the function space $C^1[-\tau, 0]$ over R^1 : For x,y in $C^1[-\tau, 0]$ and $\alpha \le \beta$ in R^1 let $(y,\beta)\ p\ (x,\alpha)$ iff $x \subset s_\alpha$, $y \subset s_\beta$ for some solution s of (1.4.2). Again it is easily verified that this relation p satisfies Definition 1.2 and hence defines a process in $C^1[-\tau, 0]$ over R^1; and that this process characterizes the original equation completely. Very similar constructions may be carried out more generally for functional-differential equation; not necessarily of retarded type, in n-space.

c) The final example concerns a one-dimensional partial differential equation

$$(1.4.3) \qquad \frac{\partial u}{\partial \theta} = f(u, \frac{\partial u}{\partial \xi}, \frac{\partial^2 u}{\partial \xi^2}, \xi, \theta)$$

with continuous $f : R^s \to R^1$; consider the corresponding homogeneous boundary value problem in the strip $\{(\xi,\theta) \in R^2 : |\xi| \le 1, \theta \ge 0$. The associated process p will act in the set P of all continuous functions on $[-1,1]$ with zero end values. For $x, y \in P$ and $\beta \ge \alpha$ in R^1 one defines that $(y,\beta)\ p\ (\omega,\alpha)$ iff $\alpha \ge 0$ and there exists a solution u of (1.4.3) with zero boundary values and such that

$$u(\xi,\alpha) = x(\xi),\ u(\xi,\beta) = y(\xi) \quad \text{for} \quad |\xi| \le 1.$$

Again, a similar construction may be carried out for higher orders, for more complicated domains and boundary conditions, and for systems of such equations.

1.5 DEFINITION. Given a process p on P over R, a relation s between P and R will be called a <u>solution</u> of p iff

(i) s is a partial map $R \to P$;

(ii) domain s is an interval in R;

(iii) $(s\beta)\ {}_\beta p_\alpha\ (s\alpha)$ holds for all $\beta \ge \alpha$ in domain s.

1.6 DEFINITION. A process p on P over R will be called solution complete iff, whenever $y_\beta p_\alpha x$, there exists a solution s of p with $s\alpha = x$, $s\beta = y$.

1.7 REMARK. If s is a solution of p, and $\alpha \in$ domain s, one may say that s is a solution through $(s\alpha,\alpha)$ or through $s\alpha$ at α. Given a process p on P over R and an $(x,\alpha) \in D$, it is natural to inquire about the set of θ's such that $y_\theta p_\alpha x$ for some $y \in P$. Some information is directly available: according to the compositivity property, such θ's constitute an interval in R containing α as left end-point, and entirely contained within the interval-component of R. This suggests the following definition.

1.8 DEFINITION. Let p be a process on P over R; define a map

$$\varepsilon : D \to R^1 \cup \{+\infty\} : \varepsilon(x,\alpha) = \sup\{\theta \in R^1 : (\theta,x,\alpha) \in E\}$$

with the supremum taken in the extended real line. The value $\varepsilon(x,\alpha)$ of the map ε at the point $(x,\alpha) \in D$ will be termed the escape time of (x,α).

1.9 REMARK. If $\alpha \leq \theta < \varepsilon(x,\alpha)$, then there exists an $y \in P$ with $y_\theta p_\alpha x$; as a partial converse, if $y_\theta p_\alpha x$ then $\alpha \leq \theta \leq \varepsilon(x,\alpha)$. In particular, $\alpha \leq \varepsilon(x,\alpha)$. Clearly, ε non-increases along p, i.e. $y_\beta p_\alpha x$ implies $\varepsilon(y,\beta) \leq \varepsilon(x,\alpha)$.

1.10 DEFINITION. Given a process p on P over R, and an $(x,\alpha) \in D$, p is said to have local existence at (x,α) iff $\alpha < \varepsilon(x,\alpha)$. Iff this obtains for all elements of D, then p is said to have local existence. (x,α) is an end-pair iff $\alpha = \varepsilon(x,\alpha)$. Process p is termed global iff $\varepsilon(x,\alpha) = +\infty$ for all $(x,\alpha) \in D$.

1.11 REMARK. For processes with local existence, one has the following version of 1.9: For given (x,α) and θ there is $y_\theta p_\alpha x$ for some $y \in P$ iff $\alpha \leq \theta < \varepsilon(x,\alpha)$.

1.12 DEFINITION. A process p on P over R is said to possess unicity iff every individual relation ${}_\beta p_\alpha$ is a partial map $P \to P$.

1.13 REMARK. Definition 1.12 is equivalent with that

$${}_\beta p_\alpha \circ ({}_\beta p_\alpha)^{-1} \subset 1 \qquad \text{for all } \alpha \leq \beta \text{ in } R.$$

An independent description of such processes will be useful. Let P be a set, $R \subset R^1$, and let there be given a partial map $t: R \times P \times R \to P$; for all $\beta \geq \alpha$ in R define partial maps

$$_{\beta}t_{\alpha} : P \to P : \ _{\beta}t_{\alpha}\ x = t\ (\beta, x, \alpha).$$

Then $_{\beta}t_{\alpha}$ are individual relations of a process in P over R (necessarily with unicity) iff the following two conditions are satisfied:

(i) $_{\alpha}t_{\alpha}\ x = x$ for all $(x,\alpha) \in P \times R$ with $(\alpha,x,\alpha) \in$ domain t,

(ii) $_{\gamma}t_{\beta} \circ {}_{\beta}t_{\alpha}x = {}_{\gamma}t_{\alpha}x$ for all $\gamma \geq \beta \geq \alpha$ in R whenever either side is defined.

In particular, this implies that $_{\theta}t_{\alpha}x$ with fixed $(x,\alpha) \in P \times R$ is defined for all $\theta \in R$ in an (possibly degenerate) interval with α as left end-point.

Next, we shall exhibit a localisation of the unicity property, in a manner similar to that employed for local existence.

1.14 DEFINITION. Let p be a process on P over R; define a map

$$\delta : D \to R^1 \cup \{+\infty\} : \delta(x,\alpha) \ = \ \sup \{\lambda \in R^1 : \alpha \leq \theta \leq \lambda,\ u_{\theta}p_{\alpha}x,\ v_{\theta}p_{\alpha}x$$

imply $u = v\}$, (the supremum is taken in the extended real line). The value $\delta(x,\alpha)$ of the map δ at the point $(x,\alpha) \in D$ will be termed the <u>extent of unicity</u> of (x,α).

1.15 LEMMA.

(i) $\alpha \leq \delta(x,\alpha) \leq +\infty$ for each $(x,\alpha) \in D$;

(ii) if $\alpha \leq \beta \leq \gamma \leq \delta\ (x,\alpha)$, then $y_{\beta}p_{\alpha}x$, $z_{\gamma}p_{\alpha}x$ imply $z_{\gamma}p_{\beta}y$;

(iii) if $y_{\beta}p_{\alpha}x$ with $\alpha \leq \beta < \delta\ (y,\beta)$ then $\delta(x,\alpha) = \delta(y,\beta)$ and $\varepsilon(x,\alpha) = \varepsilon(y,\beta)$;

(iv) δ non-decreases along p;

(v) if $\delta(x,\alpha) < +\infty$, then there exist $\theta \in R$ with arbitrarily small $\theta - \delta(x,\alpha) \geq 0$ and also u, v in P with $u \neq v$, $u_{\theta}p_{\alpha}x$, $v_{\theta}p_{\alpha}x$; in particular $\alpha \leq \delta(x,\alpha) \leq \varepsilon(x,\alpha)$;

(vi) if $\delta(x,\alpha) = \varepsilon(x,\alpha) < +\infty$ then $\varepsilon(x,\alpha) > \alpha$ and there exist $u \neq v$ in P with $u_{\theta}p_{\alpha}x$, $v_{\theta}p_{\alpha}x$. If $\alpha = \varepsilon(x,\alpha)$ then $\delta(x,\alpha) = +\infty$.

1.16 DEFINITION. Let there be given a process p on P over R and $(x,\alpha) \in D$. p is said to have <u>local unicity at</u> (x,α) iff $\alpha < \delta(x,\alpha)$. Iff this obtains for all elements of D, then p is said to have <u>local unicity</u>, p is said to have <u>global unicity at</u> (x,α) iff $\delta(x,\alpha) = +\infty$.

1.17 REMARK. Evidently p has unicity in the sense of 1.12 iff it has global unicity at all elements of D.

1.18 DEFINITION. Let p be a process on P over R; then p is said to <u>admit the period</u> τ iff $\tau \in R^1$ and

$$_{\beta-\tau}p_{\alpha-\tau} = {}_{\beta}p_{\alpha} = {}_{\beta+\tau}p_{\alpha+\tau} \quad \text{obtains for all } \beta \geq \alpha \text{ in R.}$$

1.19 REMARK. If a process p admits the period τ and s is a solution of p, then so are the partial maps s_τ and $s_{-\tau}$ obtained by "translation" $s_\tau\theta = s(\theta+\tau)$ for all $(\theta+\tau) \in$ domain s. Clearly, $\varepsilon(x,\alpha+\tau) = \varepsilon(x,\alpha) + \tau$ and $\delta(x,\alpha+\tau) = \delta(x,\alpha) + \tau$. Hence local existence, etc., obtains at $(x,\alpha+\tau)$ iff it obtains at (x,α).

1.20 DEFINITION. A process p on P over R is termed <u>stationary</u> iff it admits all periods $\theta \in R$.

1.21 REMARK. Reducing the last Definition the condition is that

$$_{\beta-\theta}p_{\alpha-\theta} = {}_{\beta}p_{\alpha} = {}_{\beta+\theta}p_{\alpha+\theta} \quad \text{obtains for all } \beta \geq \alpha \text{ and } \theta \in R.$$

Clearly $\varepsilon(x,\alpha) = \varepsilon(x,o) + \alpha$, $\delta(x,\alpha) = \delta(x,o) + \alpha$. Thus, if local existence obtains at (x,o), then it also obtains at all (x,α) for arbitrary $\alpha \in R$; and similarly for local unicity, etc. In particular, p has local existence iff $\varepsilon(x,0) > 0$ for all $(x,0) \in D$, globality iff $\varepsilon(x,0) = +\infty$ for all $(x,0) \in D$, etc.

Clearly, the individual relations of a stationary process p on P satisfy

$$_{\beta}p_{\alpha}\, x = {}_{\beta-\alpha}p_{o} \quad \text{for all } \beta \geq \alpha \text{ in R,}$$

and hence depend on a single parameter $\beta - \alpha \geq 0$. Now an independent description of such processes will be given.

Let there be given a set P, $R \subset R^1$ and a system $\{q_\alpha : \alpha \geq 0\}$ of relations q_α on P with the following two properties:

(i) $q_o \subset 1$

(ii) $q_\alpha \circ q_\beta = q_{\alpha+\beta}$ for $\alpha, \beta \geq 0$.

Then

(1.21.1) $${}_\beta p_\alpha = q_{\beta-\alpha} \quad \text{for } \beta \geq \alpha \text{ in } R$$

defines the individual relations of a stationary process in P. Conversely, given a stationary process p in P, (1.21.1) defines q_θ for $\theta \geq o$ unequivocally and then the system $\{q_\alpha : \alpha \geq 0\}$ satisfies conditions (i) and (ii).

2. Strong stability of sets with respect to an abstract process

In this section we shall discuss several conditions for strong stability, uniform-strong stability, asymptotic strong stability and uniform-asymptotic strong stability of subsets of $P \times R$ with respect to a process on P over $R \subset R^1$. We give there some theorems very similar to those on Liapunov's stability and Liapunov's functions, used in the differential equations theory. The conditions of the theorems formulated in what follows differ from the conditions of the classical theorems on Liapunov's functions in two points. The objects investigated are not solutions of a process, but the generalized trajectories (as they are used by ZUBOV, ROXIN, HALKIN, etc.) and there is supposed to be given very simple "distance-structure" on the domain of the process under discussion.

In the next section, there will be given similar results for the corresponding weak stability properties.

2.1 NOTATION. In what follows we suppose that there is given a non-void set

(2.1.1) $$m \subset P \times R$$

and a function

(2.1.2) $$g : P \times R \to R^o$$

such that $(x,\alpha) \in m$ implies $g(x,\alpha) = 0$.

2.2 DEFINITION. Let p be a process on P over R and m the set (2.1.1). The set m is said to be <u>strongly</u> <u>stable</u> iff there exists a function

(2.2.1) $$\omega : R \times R^+ \to R^+$$

such that

$$(\theta,x,\alpha) \in E \text{ with } g(x,\alpha) \leq \omega(\alpha,\eta) \text{ and } y_\theta p_\alpha x \text{ imply } g(y,\theta) \leq \eta.$$

2.3 THEOREM. Let there be given a process p on P over R, the set (2.1.1), $\delta > 0$, and a partial function $V : D \to R^o$ with the following properties:

(i) domain $V \supset \{(y,\beta) \in D : y_\beta p_\alpha x$ for some $(x,\alpha) \in D$ with $g(x,\alpha) < \delta\}$;

(ii) there are functions $a : R^o \to R^o$, $d : R \times R^+ \to R^+$, a increasing, $a(0) = 0$, such that
$(x,\alpha) \in$ domain V implies $a(g(x,\alpha)) \leq V(x,\alpha)$;
$(x,\alpha) \in$ domain V with $g(x,\alpha) \leq d(\alpha,\eta)$ implies $V(x,\alpha) \leq a(\eta)$;

(iii) V is non-increasing along p; i.e. $V(x,\alpha) \geq V(y,\beta)$ holds whenever $y_\beta p_\alpha x$; $(x,\alpha) \in$ domain V.

Then the set m is strongly stable.

Proof. Let there be given $\eta \in (0,\delta)$ and $\alpha \in R$. Set $0 < \omega(\alpha,\eta) \leq d(\alpha,\eta)$; then $g(x,\alpha) \leq \omega(\alpha,\eta)$ implies $V(x,\alpha) \leq a(\eta)$. Thus, given $(\theta,x,\alpha) \in E$ with $g(x,\alpha) \leq \omega(\alpha,\eta)$ and $y_\theta p_\alpha x$, there holds

$$a(g(y,\theta)) \leq V(y,\theta) \leq V(x,\alpha) \leq a(\eta),$$

whence, as a is increasing, there follows $g(y,\theta) \leq \eta$, what proves the strong stability of m.

2.4 DEFINITION. Let there be given a process p on P over R and the set (2.1.1). The set m is said to be <u>uniform-strongly stable</u> iff there exists an increasing function

(2.4.1) $$\omega_1 : R^+ \to R^+$$

such that

$$(\theta,x,\alpha) \in E \text{ with } g(x,\alpha) \leq \omega_1(\eta) \text{ and } y_\theta p_\alpha x \text{ imply } g(y,\theta) \leq \eta .$$

2.5 THEOREM. Let there be given a process p on P over R and the set (2.1.1). The set m is uniform-strongly stable iff there exist $\delta > 0$ and a partial function $V : D \to R^o$ with the following properties:

(i) and (iii) as in Theorem 2.3;

(ii) there are increasing functions a, b: $R^o \to R^o$, $a(0) = 0 = b(0)$, $\lim_{r \to 0_+} b(r) = 0$, such that for each $(x,\alpha) \in$ domain V there holds

$$a(g(x,\alpha)) \leq V(x,\alpha) \leq b(g(x,\alpha)).$$

Proof. First suppose that there is a function V with the properties described in the Theorem and let $\eta \in (0,\delta)$. Define $\omega_1(\eta)$ such that $0 < b(\omega_1(\eta)) \leq a(\eta)$. Given $(\theta,x,\alpha) \in E$ with $g(x,\alpha) \leq \omega_1(\eta)$ and $y_\theta p_\alpha x$, there holds

$$a(g(y,\theta)) \leq V(y,\theta) \leq V(x,\alpha) \leq b(g(x,\alpha)) \leq b(\omega_1(\eta)) \leq a(\eta),$$

so that $g(y,\theta) \leq \eta$, what proves the uniform strong stability of m. Now, let the set m be uniformly strong stable. Let ω_1 be the function (2.4.1). Set $\delta = \sup\{\omega_1(\eta) : 0 < \eta < +\infty\}$ and define a partial map V as follows:

(2.5.1) $$V : D \to R^o : V(x,\alpha) = \sup\{g(y,\theta) : y_\theta p_\alpha x\}$$

for $(x,\alpha) \in D$ with $g(x,\alpha) < \delta$. We shall prove that V satisfies the conditions (i) to (iii) of the Theorem.

Ad (i): Assume given $(x,\alpha) \in D$ with $g(x,\alpha) < \delta$. Then $g(x,\alpha) \leq \omega_1(\eta)$ for some $\eta \in R^+$, so that $g(y,\theta) \leq \eta$ for all $(y,\theta) \in D$ with $y_\theta p_\alpha x$, hence $V(x,\alpha)$ is defined.

Ad (ii): Let ω_1 be the function from (2.4.1). Denote η_1 the inverse function to ω_1. Let $(x,\alpha) \in D$ be such that $g(x,\alpha) = \omega_1(\eta)$. Then $\eta = \eta_1(g(x,\alpha))$ and according to the uniform-strong stability of m, $g(x,\alpha) = \omega_1(\eta)$ implies $g(y,\theta) \leq \eta = \eta_1(g(x,\alpha))$ whenever $y_\theta p_\alpha x$, so that according to (2.5.1) $V(x,\alpha) \leq \eta_1(g(x,\alpha))$.

Further on, $g(x,\alpha) \in \{g(y,\theta) : y_\theta p_\alpha x\}$, so that $g(x,\alpha) \leq V(x,\alpha)$.

Now, setting $a(r) = r$, $b(r) = \eta_1(r)$ for $r \in R^o$, V satisfies condition (ii).

Ad (iii): Let there be given $(\beta,x,\alpha) \in E$ with $g(x,\alpha) < \delta$ and $y_\beta p_\alpha x$. Then

$$\begin{aligned} V(y,\beta) &= \sup \{g(z,\theta) : z_\theta p_\beta y\} \\ &\leq \sup \{g(z,\theta) : z_\theta p_\beta o_\beta p_\alpha x\} \\ &= \sup \{g(z,\theta) : z_\theta p_\alpha x\} = V(x,\alpha). \end{aligned}$$

2.6 THEOREM. Let there be given a process p on P over R admitting the period $\tau > 0$ and the set (2.1.1) such that the map (2.1.2) is periodic with respect to α with the period τ. If the set m is strongly stable, and there is a function $\mu : R^+ \to R^+$ satisfying the relation $\omega(\alpha,\eta) \geq \mu(\eta)$ for all $\alpha \in [0,\tau]$ and $\eta \in R^+$, then it is uniform-strongly stable.

Proof. Since m is strongly stable, then $(\theta,x,\alpha) \in E$ with $g(x,\alpha) \leq \omega(\alpha,\eta)$ and $y_\theta p_\alpha x$ imply $g(y,\theta) \leq \eta$. Set $\omega_1(\eta) = \mu(\eta)$. Then

(2.6.1) $(\theta,x,\alpha) \in E$ with $\alpha \in [0,\tau]$, $g(x,\alpha) \leq \omega_1(\eta)$ and $y_\theta p_\alpha x$ imply $g(y,\theta) \leq \eta$.

Now, let there be given $(\theta,x,\alpha) \in E$, $\alpha = \alpha' + k\tau$, $\alpha' \in [0,\tau)$, k integer, $y_\theta p_\alpha x$. Then $y_{\theta-k\tau} p_{\alpha'} x$, whence, according to (2.6.1), there follows $g(y,\theta-k\tau) \leq \eta$, so that

$(\theta,x,\alpha) \in E$ with $g(x,\alpha) \leq \omega_1(\eta)$ and $y_\theta p_\alpha x$ imply
$g(y,\theta) = g(y,\theta-k\tau) \leq \eta$,

what proves the uniform-strong stability of m.

2.7 THEOREM. Let the assumptions of Theorem 2.6 be satisfied. Then there is a partial function V with the properties 2.5(i) to 2.5(iii), periodic with respect to α with the period τ.

Proof. According to Theorem 2.6 m is uniform-strongly stable, so that the existence of a partial function V with the properties described in Theorem 2.5 follows from this Theorem. The periodicity of V follows from the relations $g(y,\theta) = g(y,\theta + \tau)$ and $y_\theta p_\alpha x$ iff $y_{\theta+\tau} p_{\alpha+\tau} x$, since then

$$V(x,\alpha) = \sup \{g(y,\theta) : y_\theta p_\alpha x\}$$
$$= \sup \{g(y,\theta + \tau) : y_{\theta+\tau} p_{\alpha+\tau} x\} = V(x,\alpha + \tau).$$

2.8 DEFINITION. Let there be given a process p on P over R and the set (2.1.1). The set m is said to be <u>asymptotically strongly-stable</u> iff it is strongly stable and there exist functions

$$\omega_2 : R \to R^+$$

and

$$T_1 : D \times R^+ \to R^+$$

such that $(\theta,x,\alpha) \in E$ with $g(x,\alpha) \le \omega_2(\alpha)$ and $y_\theta p_\alpha x$ with $\theta \ge \alpha + T_1(x,\alpha,\eta)$ imply $g(y,\theta) \le \eta$.

2.9 THEOREM. Let there be given a global solution complete process p, the set (2.1.1), $\delta > 0$, and a partial function $V : D \to R^o$ with the following properties:

(i) and (ii) as in Theorem 2.3;

(iii) there is an increasing function $c : R^o \to R^o$, $c(0) = 0$, such that $V(y,\theta) - V(x,\alpha) \le - \int_\alpha^\theta c(V(s\sigma,\sigma))d\sigma$ for each $(x,\alpha) \in$ domain V and each solution s through (x,α) with $s\theta = y$.

Then m is asymptotically strongly stable.

Proof. From (iii) there follows $V(y,\theta) \le V(\omega,\alpha)$ whenever $y_\theta p_\alpha x$, so that Theorem 2.3 takes place and m is strongly-stable. Hence, corresponding to each $\alpha \in R$ and δ from (i), there is $\omega(\alpha,\delta)$ such that $(\theta,x,\alpha) \in E$ with $g(x,\alpha) \le \omega(\alpha,\delta)$ and $y_\theta p_\alpha x$ imply $g(y,\theta) \le \delta$. Set $\omega_2(\alpha) = \omega(\alpha,\delta)$ and take $(x,\alpha) \in$ domain V with $g(x,\alpha) \le \omega_2(\alpha)$. Denote $S(x,\alpha)$ the family of all solutions of p through (x,α) with $[\alpha,+\infty) \subset$ domain s. Since V is non-increasing and non-negative, then corresponding to each $s \in S(x,\alpha)$ there exists a limit

(2.10.1) $$V(x,\alpha) \ge V_o(s) = \lim_{\sigma\to+\infty} V(s\sigma,\sigma) \ge 0.$$

Suppose that there is an $s \in S(x,\alpha)$ with $V_0(s) > 0$. Then

$$V(s\theta,\theta) \leq V(x,\alpha) - \int_\alpha^\theta c(V_0(s))d\sigma \leq V(x,\alpha) - c(V_0(s))(\theta-\alpha).$$

Setting $\theta > \alpha + \frac{V(x,\alpha)}{c(V_0(s))}$, there follows $V(s\theta,\theta) < 0$, what contradicts (ii). Thus, for each $s \in S(x,\alpha)$ we have $V_0(s) = 0$, and since p is solution complete, then $y_\theta p_\alpha x$ implies $y = s\theta$ for some $s \in S(x,\alpha)$, so that from (2.9.1) and (ii) there follows the existence of $T'(x,\alpha,\eta')$ such that $a(g(y,\theta)) \leq \eta'$ for each $\theta \geq \alpha + T'(x,\alpha,\eta')$. Setting $T_1(x,\alpha,\eta) = T'(x,\alpha,a(\eta))$, one has $g(y,\theta) \leq \eta$ for $\theta \geq \alpha + T_1(x,\alpha,\eta)$, what proves the Theorem.

2.10 DEFINITION. Let there be given a process p on P over R and the set (2.1.1). The set m is said to be <u>uniform-asymptotically strongly</u> stable iff there are a number $\Omega > 0$ and functions ω_3, $T : R^+ \to R^+$ with the following properties:

(i) $(\theta,x,\alpha) \in E$ with $g(x,\alpha) \leq \omega_3(\eta)$ and $y_\theta p_\alpha x$ imply $g(y,\theta) \leq \eta$;

(ii) $(\theta,x,\alpha) \in E$ with $g(x,\alpha) \leq \Omega$ and $y_\theta p_\alpha x$ with $\theta \geq \alpha + T(\eta)$ imply $g(y,\theta) \leq \eta$.

2.11 THEOREM. Let there be given a global solution complete process p on P over R, the set (2.1.1), $\delta > 0$, and a partial function $V : D \to R^0$ with the following properties:

(i) and (ii) as in Theorem 2.5;

(iii) there is an increasing function $c : R^0 \to R^0$, $c(0) = 0$, such that $V(y,\theta) - V(x,\alpha) \leq -\int_\alpha^\theta c(g(s\sigma,\sigma))d\sigma$ holds for each $(x,\alpha) \in$ domain V and each solution s through (x,α) with $s\theta = y$.

Then m is uniform-asymptotically strongly stable.

Proof. Let $0 < \Omega < \delta$. To prove 2.10(i), given $\eta \in (0,\delta)$, let $\omega_3(\eta) > 0$ be such that $b(\omega_3(\eta)) \leq a(\eta)$. Then $(\theta,x,\alpha) \in E$ with $g(x,\alpha) \leq \omega_3(\eta)$ and $y_\theta p_\alpha x$ imply $a(g(y,\theta)) \leq V(y,\theta) \leq V(x,\alpha) \leq b(g(x,\alpha)) \leq b(\omega_3(\eta)) = a(\eta)$, whence there follows $g(y,\theta) \leq \eta$, so that 2.10(i) holds.

To prove 2.10(ii), given $\eta \in (0,\delta)$, define $T(\eta) = \frac{b(\Omega)}{c(\omega_3(\eta))}$. Let there be given $(x,\alpha) \in D$ with $g(x,\alpha) \leq \Omega$. To finish the proof, it suffices to show that there is (z,β) with $z_\beta p_\alpha x$, $\beta \in [\alpha,\alpha + T(\eta)]$, such that $g(z,\beta) \leq \omega_3(\eta)$; since then, using the preceding part of the proof there follows directly $g(y,\theta) \leq \eta$ for all (y,θ) with $y_\theta p_\alpha x$ and $\theta \geq \alpha + T(\eta) > \beta$.

Let us suppose that $g(z,\beta) > \omega_3(\eta) > 0$ holds for each (z,β) with $z_\beta p_\alpha x$ and $\alpha \leq \beta \leq \alpha + T(\eta)$. Hence, for $\beta = \alpha + T(\eta)$ there holds $0 < a(g(z,\beta)) \leq V(z,\beta) \leq V(x,\alpha) - c(\omega_3(\eta)).(\beta-\alpha) \leq b(g(x,\alpha)) - c(\omega_3(\eta))(\beta-\alpha) \leq b(\Omega) - c(\omega_3(\eta)).T(\eta) = 0$, and this contradiction finishes the proof.

2.12 THEOREM. Let there be given a global solution complete process p on P over R admitting the period $\tau > 0$, the set (2.1.1), and a partial function $V : D \to R^0$ with the following properties:

(i) to (iii) as in Theorem 2.9;

(iv) there is $\lambda > 0$ such that $V(x,\alpha) \leq \lambda$ for all $(x,\alpha) \in$ domain V with $g(x,\alpha) \leq \delta$ and $\alpha \in [0,\tau]$;

(v) there is a positive function $\mu : R^+ \to R^+$ such that the function (2.2.1) satisfies the relation $\omega(\alpha,\eta) \geq \mu(\eta)$ for each $\alpha \in [0,\tau]$ and $\eta \in R^+$.

Then m is uniform-asymptotically strongly stable.

Proof. According to Theorem 2.6, m is uniform-strongly stable, so that 2.10(i) is satisfied.

Now, given $\eta \in (0,\delta)$ take $0 < \Omega < \delta$ and suppose that there does not exist $T(\eta)$, such that 2.10(ii) holds. Then there are sequences $(\theta_n,x_n,\alpha_n) \in E$, $y_n\, {}_{\theta_n}p_{\alpha_n}\, x_n$, such that $g(x_n,\alpha_n) \leq \Omega$, $\alpha_n \in [0,\tau]$, $\theta_n \geq \alpha_n + n$, $g(y_n,\theta_n) > \eta$. Let s^n be a solution through (x_n,α_n) with $s^n\theta_n = y_n$. Then there holds

$$0 < a(\eta) < a(g(y_n,\theta_n)) \leq V(y_n,\theta_n) \leq V(x_n,\alpha_n) - \int_{\alpha_n}^{\theta_n} c(V(s^n\sigma,\sigma))d\sigma \leq \lambda - c(V(y_n,\theta_n))$$

$$(\theta_n - \alpha_n) \leq \lambda - c(a(\eta)).n,$$

whence, for $n \geq \frac{\lambda}{c(a(\eta))}$, there follows $0 < \lambda - c(a(\eta)) \frac{\lambda}{c(a(\eta))} = 0$, and this contradiction finishes the proof.

3. Weak stability of sets with respect to an abstract process

3.1 DEFINITION. Let there be given a process p on P over R and the set (2.1.1). The set m is said to be weakly stable (with respect to p) iff there exists a function

$$\omega : R \times R^+ \to R^+$$

such that $(\theta,x,\alpha) \in E$ with $g(x,\alpha) \leq \omega(\alpha,\eta)$ imply $g(y,\theta) \leq \eta$ for some $y \in {}_\theta p_\alpha x$.

3.2 THEOREM. Let there be given a process p on P over R, the set (2.1.1), $\delta > 0$, and a partial function $V : D \to R^0$ with the following properties:

(i) and (ii) as in Theorem 2.3;

(iii) corresponding to each $(\theta,x,\alpha) \in E$ with $(x,\alpha) \in$ domain V there is $y \in {}_\theta p_\alpha x$ such that $V(y,\theta) \leq V(x,\alpha)$.

Then m is weakly stable.

Proof. Let there be given $\eta \in (0,\delta)$ and $\alpha \in R$. Set $0 < \omega(\alpha,\eta) \leq d(\alpha,\eta)$. Then, according to (ii), $(x,\alpha) \in D$ with $g(x,\alpha) \leq \omega(\alpha,\eta)$ imply $V(x,\alpha) \leq a(\eta)$. Furthermore, given $(\theta,x,\alpha) \in E$ with $g(x,\alpha) \leq \omega(\alpha,\eta)$, there exists $y \in {}_\theta p_\alpha x$ such that $V(y,\theta) \leq V(x,\alpha)$, so that then $a(g(y,\theta) \leq V(y,\theta) \leq V(x,\alpha) \leq a(\eta)$, whence $g(y,\theta) \leq \eta$ proving Theorem 3.2.

3.3 DEFINITION. Let there be given a process p on P over R and the set (2.1.1). The set m is said to be uniform-weakly stable iff there exists an increasing function

$$\omega_1 : R^+ \to R^+$$

such that $(\theta,x,\alpha) \in E$ with $g(x,\alpha) \leq \omega_1(\eta)$ imply $g(y,\theta) \leq \eta$ for some $y \in {}_\theta p_\alpha x$.

3.4 THEOREM. Let there be given a process p on P over R, the set (2.1.1), $\delta > 0$, and a partial function $V : D \to R^0$ with the following properties:

(i) and (iii) as in Theorem 3.3;

(ii) there are increasing functions a, b : $R^0 \to R^0$, $a(0) = 0 = b(0)$, $\lim_{r \to 0_+} b(r) = 0$, such that $a(g(x,\alpha)) \leq V(x,\alpha) \leq b(g(x,\alpha))$ holds for each $(x,\alpha) \in$ domain V.

Then m is uniform-weakly stable.

Proof. Given $\eta \in (0,\delta)$, define $\omega_1(\eta)$ so that $0 < b(\omega_1(\eta)) \le a(\eta)$. Then, corresponding to $(\theta,x,\alpha) \in E$ with $g(x,\alpha) \le \omega_1(\eta)$ there is $y \in {}_\theta p_\alpha x$ such that $V(y,\theta) \le V(x,\alpha)$, whence

$$a(g(y,\theta)) \le V(y,\theta) \le V(x,\alpha) \le b(g(x,\alpha)) \le b(\omega_1(\eta)) \le a(\eta)$$

implies $g(y,\theta) \le \eta$, and this proves the theorem.

3.5 THEOREM. Let there be given a process p on P over R admitting the period $\tau > 0$, and the set (2.1.1) such that the corresponding map (2.1.2) is periodic with respect to α with the period τ. If m is weakly stable, and there exists a function $\mu : R^+ \to R^+$ such that $\omega(\alpha,\eta) \ge \mu(\eta)$ for all $\alpha \in [0,\tau]$ and $\eta \in R^+$, then it is uniform-weakly stable.

Proof. Since m is weakly stable, there is a map ω such that $(\theta,x,\alpha) \in E$ with $g(x,\alpha) \le \omega(\alpha,\eta)$ imply $g(y,\theta) \le \eta$ for some $y \in {}_\theta p_\alpha x$. Set $\omega_1(\eta) = \mu(\eta)$. Then

(3.5.1) $(\theta,x,\alpha) \in E$ with $g(x,\alpha) \le \omega_1(\eta)$, $\alpha \in [0,\tau]$ imply
$g(y,\theta) \le \eta$ for some $y \in {}_\theta p_\alpha x$.

Now, let $\alpha \in R$ be arbitrary, $\alpha = \alpha' + k\tau$ with $\alpha' \in [0,\tau)$ and k integer. Let $(\theta,x,\alpha) \in E$ with $g(x,\alpha) \le \omega_1(\eta)$. Then $(\theta - k\tau, x, \alpha') \in E$ and according to (3.5.1) there is $y \in {}_{\theta-k\tau}p_{\alpha'}x$ with $g(y,\theta - k\tau) \le \eta$. Since $y\,{}_{\theta-k\tau}p_{\alpha'}\,x$ iff $y\,{}_\theta p_\alpha x$, there holds $y \in {}_\theta p_\alpha x$ and $g(y,\theta) = g(y,\theta - k\tau) \le \eta$, what proves the uniform-weak stability of m.

3.6 DEFINITION. Let there be given a process p and the set (2.1.1). The set m is said to be <u>asymptotically</u> <u>weakly</u> <u>stable</u> iff it is weakly stable and there exist functions

$$\omega_2 : R \to R^+, \; T_1 : D \times R^+ \to R^+$$

such that $(\theta,x,\alpha) \in E$ with $g(x,\alpha) \le \omega_2(\alpha)$ and $\theta \ge \alpha + T_1(\alpha,\eta)$ imply $g(y,\theta) \le \eta$ for some $y \in {}_\theta p_\alpha x$.

3.7 THEOREM. Let there be given a global process p on P over R, the set (2.1.1), $\delta > 0$, and a partial function $V : D \to R^0$ with the following properties:

(i) and (iii) as in Theorem 3.2;

(ii) there is an increasing function $c : R^0 \to R^0$, $c(0) = 0$, such that corresponding to each $(\beta,x,\alpha) \in E$ with $g(x,\alpha) < \delta$ there is a solution s^β through (x,α) such that there holds

(3.7.1) $$V(s^\beta\theta,\theta) - V(x,\alpha) \le -\int_\alpha^\theta c(V(s^\beta\sigma,\sigma))d\sigma \text{ for each } \theta \in [\alpha,\beta].$$

Then m is asymptotically weakly stable.

Proof. From (ii) there follows $V(y,\theta) \le V(x,\alpha)$ for each $(\theta,x,\alpha) \in E$ and some $y \in {}_\theta p_\alpha x$, so that Theorem 3.2 takes place; hence m is weakly stable.

Set $\omega_2(\alpha) = \omega(\alpha,\delta)$ and take $(x,\alpha) \in D$ with $g(x,\alpha) \le \omega_2(\alpha)$. Then, corresponding to each $\beta > \alpha$ there is a solution s^β such that (3.7.1) holds. Moreover,

(3.7.2) $$V(s^\beta\beta,\beta) \le V(x,\alpha) - c(V(s^\beta\beta,\beta))(\beta-\alpha).$$

Now, it suffices to prove that there is $T(\alpha,\eta)$ such that $V(s^\beta\beta,\beta) \le a(\eta)$ for each $\beta \ge \alpha + T(\alpha,\eta)$. Suppose that this is not true. Then there are arbitrary large β's such that $V(s^\beta\beta,\beta) > a(\eta)$. Hence and from (3.7.2) there follows

$$V(s^\beta\beta,\beta) \le V(x,\alpha) - c(a(\eta))(\beta-\alpha),$$

so that for $\beta > \alpha + \dfrac{V(x,\alpha)}{c(a(\eta))}$ there follows $V(s^\beta\beta,\beta) < 0$, what contradicts the non-negativeness of V and finishes the proof.

3.8 DEFINITION. Let there be given a process p on P over R and the set (2.1.1). The set m is said to be uniform-asymptotically weakly stable iff there exist a number $\Omega > 0$ and functions $\omega_3, T : R^+ \to R^+$ with the following properties:

(i) $(\theta,x,\alpha) \in E$ with $g(x,\alpha) \le \omega_3(\eta)$ imply $g(y,\theta) \le \eta$ for some $y \in {}_\theta p_\alpha x$;

(ii) $(\theta,x,\alpha) \in E$ with $g(x,\alpha) \le \Omega$ and $\theta \ge \alpha + T(\eta)$ imply $g(y,\theta) \le \eta$ for some $y \in {}_\theta p_\alpha x$.

3.9 THEOREM. Let there be given a global process p on P over R, $\delta > 0$, the set (2.1.1), and a partial function $V : D \to R^0$ with the following properties:

(i) and (ii) as in Theorem 3.5;

(iii) there is an increasing function $c : R^0 \to R^0$, $c(0) = 0$, such that corresponding to each $(\beta,x,\alpha) \in E$ with $g(x,\alpha) < \delta$ there is a solution s^β through (x,α)

such that $V(s^{\beta}\theta,\theta) - V(x,\alpha) \leq - \int_{\alpha}^{\theta} c(g(s^{\beta}\sigma,\sigma))d\sigma$ holds for each $\theta \in [\alpha,\beta]$.

Then m is uniform-asymptotically weakly stable.

Proof. Set $0 < \Omega < \delta$ and given $\eta \in (0,\delta)$, define $\omega_3(\eta)$ so that $0 < b(\omega_3(\eta)) \leq a(\eta)$ and $T(\eta) = \dfrac{b(\Omega)}{c(\omega_3(\eta))}$. Then, corresponding to $(\beta,x,\alpha) \in E$ with $g(x,\alpha) \leq \omega_3(\eta)$, there exists a solution s^{β} through (x,α) such that

$$a(g(s^{\beta}\beta,\beta) \leq V(s^{\beta}\beta,\beta) \leq V(x,\alpha) \leq b(g(x,\alpha)) \leq b(\omega_3(\eta)) \leq a(\eta),$$

whence there follows $g(s^{\beta}\beta,\beta) \leq \eta$, and 3.8(i) holds.

Now, let there be given $(x,\alpha) \in D$ with $g(x,\alpha) \leq \Omega$. To finish the proof, it is sufficient to show that corresponding to each $\beta \geq \alpha + T(\eta)$ there is $\theta \in [\alpha,\alpha + T(\eta)]$ such that $g(s^{\beta}\theta,\theta) \leq \omega_3(\eta)$; since hence, using the preceding part of the proof, there follows directly $g(s^{\beta}\beta,\beta) \leq \eta$.

Let us suppose that $g(s^{\beta}\theta,\theta) > \omega_3(\eta)$ holds for each $\theta \in [\alpha,\alpha + T(\eta)]$. Then $0 < a(g(s^{\beta}\theta,\theta)) \leq V(s^{\beta}\theta,\theta) \leq V(x,\alpha) - c(\omega_3(\eta))(\theta-\alpha) \leq b(g(x,\alpha)) - c(\omega_3(\eta))(\theta-\alpha)$, whence especially for $\theta = \alpha + T(\eta)$ there follows

$$0 < b(\Omega) - c(\omega_3(\eta)) \frac{b(\Omega)}{c(\omega_3(\eta))} = 0,$$

and this contradiction finishes the proof.

3.10 THEOREM. Let there be given a global process p on P over R admitting the period $\tau > 0$, the set (2.1.1), $\delta > 0$, and a partial function $V : D \to R^{o}$ with the following properties:

(i) to (iii) as in Theorem 3.7;

(iv) there is $\lambda > 0$ such that $V(x,\alpha) \leq \lambda$ for each $(x,\alpha) \in$ domain V with $g(x,\alpha) < \delta$ and $0 \leq \alpha \leq \tau$;

(v) there is a positive function $\mu : R^{+} \to R^{+}$ such that the function (3.1.1) satisfies the relation $\omega(\alpha,\eta) \geq \mu(\eta)$ for each $\alpha \in [0,\tau]$ and $\eta \in R^{+}$.

Then m is uniform-asymptotically weakly stable.

Proof. According to Theorem 3.6 m is uniform-weakly stable, so that there is a function $\omega_3 : R^+ \to R^+$ such that $(\theta,x,\alpha) \in E$ with $g(x,\alpha) \leq \omega_3(\eta)$ imply $g(y,\theta) \leq \eta$ for some $y \in {}_\theta p_\alpha x$, and thus 3.8(i) obtains.

Now, let $\Omega \in (0,\delta)$ be arbitrary. Given $\eta \in (0,\delta)$, suppose that there does not exist $T(\eta)$ such that 3.8(ii) holds. Then, corresponding to each positive integer n, there is $(\theta_n, x_n, \alpha_n)$ with $g(x_n,\alpha_n) \leq \Omega$, $0 \leq \alpha_n \leq \tau$, $\theta_n \geq \alpha_n + n$, and $g(y,\theta_n) > \eta$ for each $y \in {}_{\theta_n} p_{\alpha_n} x_n$. Let for each n s^{θ_n} be the solution through (x_n,α_n) satisfying (ii). Denote $y_n = s^{\theta_n}{}_{\theta_n}$. Then $g(y_n,\theta_n) > \eta$ and for $n > \dfrac{\lambda}{c(a(\eta))}$ there holds $0 < a(\eta) \leq a(g(y_n,\theta_n)) \leq V(y_n,\theta_n) \leq V(x_n,\alpha_n) - \int_{\alpha_n}^{\theta_n} c(V(s^{\theta_n}{}_\sigma,\sigma))d\sigma \leq \lambda - c(V(y_n,\theta_n))(\theta_n - \alpha_n) \leq \lambda - c(a(\eta))(\theta_n - \alpha_n) \leq \lambda - c(a(\eta)).\, n \leq 0$, and this proves the existence of $T(\eta)$.

4. Some "hereditary" stability properties of sets

4.1 NOTATION. In Theorems of this last paragraph it is assumed that there are given a process p on P over R with the set m from (2.1.1) and the corresponding map g from (2.1.2), as well as a process p^0 on $R^0 = \langle 0, +\infty)$ over R with the set $m^0 = \{(0,\theta) \in R^0 \times R : -\infty < \theta < +\infty\}$ and the corresponding map $g^0 : R^0 \times R \to R^0 : g^0(r,\theta) = r$. These assumptions will not be repeated. Before setting up the theorems we shall formulate the following two conditions.

4.2 CONDITIONS. There are partial functions $V : P \times R \to R$, $a, b : R^0 \to R^0$, $a(0) = b(0) = 0$, a, b increasing, $\lim_{r \to +\infty} a(r) = +\infty$, $\lim_{r \to 0_+} b(r) = 0$,

with the following properties:

(i) given $(\theta,x,\alpha) \in E$, $r \in R^+$ with $V(x,\alpha) \leqq r$, and $y \in {}_\theta p_\alpha x$, there exists $s \in {}_\theta p_\alpha{}^0 r$ such that $V(y,\theta) \leqq s$;

(ii) given $(\theta,x,\alpha) \in E$. $r \in R^+$ with $V(x,\alpha) \leqq r$, there exists $s \in {}_\theta p_\alpha{}^0 r$ such that $V(y,\theta) \leqq s$ for some $y \in {}_\theta p_\alpha x$;

(iii) $a(g(x,\alpha)) \leqq V(x,\alpha) \leqq b(g(x,\alpha))$ holds for each $(x,\alpha) \in$ domain V.

4.3 THEOREM. Let conditions 4.2(i) and (iii) be satisfied. If m^o is strongly stable with respect to p^o, then m is strongly stable with respect to p.

Proof. Accordingly to the assumption, there is a positive function $\omega^o(\alpha,\eta^o)$ such that $s \in {}_\theta p_\alpha{}^o\omega^o(\alpha,\eta^o)$ implies $s \leqq \eta^o$. Given $\alpha \in R$, $\eta \in R^+$, define $\omega(\alpha,\eta)$ so that $0 < b(\omega(\alpha,\eta)) \leqq \omega^o(\alpha,a(\eta))$, and set $\eta^o = a(\eta)$. Then, given $(\theta,x,\alpha) \in E$ with $g(x,\alpha) \leqq \omega(\alpha,\eta)$ and $y \in {}_\theta p_\alpha x$, there holds

$$V(x,\alpha) \leqq b(g(x,\alpha)) \leqq b(\omega(\alpha,\eta)) \leqq \omega^o(\alpha,\eta^o),$$

so that there exists $s \in {}_\theta p_\alpha{}^o\omega^o(\alpha,\eta^o)$ with $V(y,\theta) \leqq s$. Hence, using 4.1(iii), $a(g(y,\theta)) \leqq V(y,\theta) = s \leqq \eta^o = a(\eta)$, whence $g(y,\theta) \leqq \eta$.

4.4 THEOREM. Let conditions 4.1(ii) and (iii) be satisfied. If m^o is weakly stable with respect to p^o, then m is weakly stable with respect to p. Proof is an easy modification of the proof of the preceding theorem.

4.5 THEOREM. Let conditions 4.1(i) and (iii) be satisfied. If m^o is uniform-strongly stable with respect to p^o, then m is uniform-strongly stable with respect to p. Proof is the same as that of Theorem 4.3 with $\omega^o(\alpha,\eta^o)$ and $\omega(\alpha,\eta)$ replaced by $\omega_1^o(\eta^o)$ and $\omega_1(\eta)$.

4.6 THEOREM. Let conditions 4.1(ii) and (iii) be satisfied. If m^o is uniform-weakly stable with respect to p^o, then m is uniform-weakly stable with respect to p.

4.7 THEOREM. Let conditions 4.1(i) and (iii) be satisfied. If m^o is asymptotically strongly stable with respect to p^o, then m is asymptotically strongly stable with respect to p.

Proof. According to the assumptions there are functions $\omega^o(\alpha,\eta^o)$, $\omega_2^o(\alpha)$ and $T_1^o(\alpha,\eta^o)$ such that $s \in {}_\theta p_\alpha{}^o\omega^o(\alpha,\eta^o)$ implies $s \leqq \eta^o$, and $t \in {}_\theta p_\alpha{}^o\omega_2^o(\alpha)$ with $\theta \geqq \alpha + T_1^o(\alpha,\eta^o)$ imply $t \leqq \eta^o$. Given $\alpha \in R$, $\eta \in R^+$, define $\omega(\alpha,\eta)$, $\omega_2(\alpha)$ and $T_1(\alpha,\eta)$ to satisfy the relations

$$0 < b(\omega(\alpha,\eta)) \leqq \omega^o(\alpha,a(\eta)),$$

$$0 < b(\omega_2(\alpha)) \leqq \omega_2^o(\alpha),$$

$$T_1(\alpha,\eta) = T_1^o(\alpha,a(\eta)).$$

In the same way as in the proof of Theorem 4.3 there may be proved that $\omega(\alpha,\eta)$ so defined, satisfies the conditions of the strong stability of m. To prove remaining part of the theorem, given $\alpha \in R$ and $\eta \in R^+$, set $\eta^o = a(\eta)$. Let there be given $(\theta,x,\alpha) \in E$ with $g(x,\alpha) \leqq \omega_2(\alpha)$, $\theta \geqq \alpha + T_1(\alpha,\eta) = \alpha + T_1^o(\alpha,\eta^o)$ and $y \in {}_\theta p_\alpha x$. Then there holds

$$V(x,\alpha) \leqq b(g(x,\alpha)) \leqq b(\omega_2(\alpha)) \leqq \omega_2^o(\alpha),$$

so that there exists $t \in {}_\theta p_\alpha{}^o \omega_2^o(\alpha)$ such that

$$V(y,\theta) \leqq t \leqq \eta^o = a(\eta),$$

whence there follows

$$a(g(y,\theta)) \leqq V(y,\theta) \leqq a(\eta),$$

so that $g(y,\theta) \leqq \eta$, what proves that $\omega_2(\alpha)$ and $T_1\ (\alpha,\eta)$ satisfy conditions of Definition 2.9, and so finishes the proof.

4.8 THEOREM. Let conditions 4.1(ii) to (iii) be satisfied. If m^o is asymptotically weakly stable with respect to p^o, then m is asymptotically weakly stable with respect to p.

4.9 THEOREM. Let conditions 4.1(i) and (iii) be satisfied. If m^o is uniform-asymptotically strongly stable with respect to p^o, then m is uniform-asymptotically strongly stable with respect to p.

4.10 THEOREM. Let conditions 4.1(ii) and (iii) be satisfied. If m^o is uniform-asymptotically weakly stable with respect to p^o, then m is uniform-asymptotically weakly stable with respect to p.

REFERENCES

[1] AUSLANDER, J.; BHATIA, N. P.; SEIBERT, P.,
Attractors in dynamical systems. Bol. Soc. Mat. Mex. 9, pp. 55-66.(1964).

[2] AUSLANDER, J.; SEIBERT, P.,
Prolongations and generalized Liapunov functions, Int. Symposium on Nonlinear Differential Equations and Nonlinear Mechanics. Academic Press, New York - London (1963).

[3] BHATIA, N. P.,
Weak attractors in dynamical systems. Bol. Soc. Mat. Mex. 11, pp. 56-64 (1966).

[4] BHATIA, N. P.,
Stability and Liapunov functions in dynamical systems. Contrib. Diff. Eqs. 3, No. 2, pp. 175-188 (1964).

[5] BHATIA, N. P.; SZEGÖ, G. P.,
Weak attractors in R^n, Technical Note BN 465 of the Institute for Fluid Dynamics and Applied Mathematics, University of Maryland (1966).

[6] GOTTSCHALK, W. H.; HEDLUND, G. A.,
Topological Dynamics. A.M.S. Colloq. Publ. Vol. XXXVI, Providence, R.I. (1955).

[7] HÁJEK, O.,
Dynamical Systems in the Plane. Academic Press New York - London (1968).

[8] HÁJEK, O.,
Differentiable representation of flows. Commentationes Mathematicae Universitatis Carolinae 7, No. 2, pp. 219-225 (1966).

[9] HÁJEK, O.,
Axiomatization of differential equation theory. To be published in the Proceedings of the 2nd EQUADIFF Conference held in Bratislava in September 1966.

[10] HÁJEK, O.,
Theory of processes I. Czech. Math. Journ. 17 (92), No. 2, pp. 159-199 (1967).

[11] HALKIN, H.,
Topological aspects of optimal control of dynamical polysystems. Contrib. Diff. Eqs. 3, No. 4, pp. 377-385 (1964).

[12] KURZWEIL, J.,
Exponentially stable integral manifolds, averaging principle and continuous dependence on a parameter. Czech. Math. Journ. 16 (91), No. 3, pp. 380-423; No. 4, pp. 463-492 (1966).

[13] KURZWEIL, J.,
Invariant manifolds for flows. Proceedings of the Symposium on Dynamical Systems and Differential Equations held in Mayaguez, Puerto Rico (1965).

[14] NAGY, J.,
Lyapunov's direct method in abstract local semi-flows. Commentationes Mathematicae Universitatis Carolinae 8, No. 2, pp. 257-266 (1967).

[15] NAGY, J.,
Stability of sets with respect to continuous local semi-flows. To appear in CMUC.

[16] NAGY, J.,
Stability in continuous local semi-flows. Časopis pro pěstování matematiky 93, No. 1, pp. 8-21 (1968).

[17] NEMYCKII, V. V.; STEPANOV, V. V.,
Qualitative Theory of Differential Equations. 2nd Ed. Gostechizdat, Moscow - Leningrad (1949); English transl. Princeton Univ. Press, Princeton, N.J. (1960).

[18] ROXIN, E.,
On generalized dynamical systems defined by contingent equations. Journ. Diff. Eqs. 1, No. 2, pp. 188-205 (1965)

[19] ROXIN, E.,
Local definition of generalized control systems. Michig. Math. Journ. 13, pp. 91-96 (1966).

[20] ROXIN, E.,
Stability in general control systems. J. Diff. Eqs. 1, pp. 115-150 (1965).

[21] ROXIN, E.,
On stability in control systems. J. SIAM Control 3, No. 3, pp. 357-372 (1966).

[22] SEIBERT, P.,
Stability under perturbations in generalized dynamical systems. Int. Symposium on Nonlinear Differential Equations and Nonlinear Mechanics. Academic Press, New York - London (1963).

[23] URA, T.,
Sur le courant extérieur à une région invariante. Funk. Ekv. 2, pp. 143-200 (1959).

[24] URA, T.,
On the flow outside a closed invariant set; stability, relative stability and saddle sets. Contrib. Diff. Eqs. 3, No. 3, pp. 249-294 (1964).

[25] ZUBOV, V. I.,
Methods of A. M. Lyapunov and Their Application. Izdat. Leningrad. Univ., Leningrad (1957); English transl. P. Nordhoff Ltd., Groningen (1964).

INVARIANCE FOR CONTINGENT EQUATIONS *)

James A. Yorke

Let U be an open subset of $R \times R^d$, and let $f : U \longrightarrow R^d$ be continuous. For $(t,x) \in U$ let $\Omega(t,x) \subset R^d$ be a non-empty closed set. We say Ω is compactly upper semi-continuous if for each compact $Q \subset U$, $\{(v,t,x) : v \in \Omega(t,x) \text{ and } (t,x) \in U\} \subset R^d \times U$ is compact. We are interested in

(E) $$\dot{x} = f(t,x) \ ,$$

(C) $$\dot{x} \in \Omega(t,x) \ .$$

For any curve ϕ, we will let D_ϕ denote the interval that is its domain. By a solution of (C) we will mean an absolutely continuous function ϕ defined on D_ϕ such that $\frac{d}{dt}\phi(t) \in \Omega(t,\phi(t))$, a.e. D_ϕ. We will also assume that each "solution" ϕ of (E) or (C) is defined on as large a domain as possible; i.e. D_ϕ cannot be extended to a larger interval. Any (local) solution of (E) or (C) can be extended to such a "maximal solution" though there does not have to be a unique extension.

A set $W \subset U$ is weakly invariant (for (E) or (C)) if it is the union of the graphs of some set of solutions (of (E) or (C), respectively); that is, for ach $(t,x) \in W$, there is a solution ϕ with $\phi(t) = x$ such that $(\tau,\phi(\tau)) \in W$ for all $\tau \in D_\phi$. W is positively weakly invariant if the above holds for all $\tau \in D_\phi \cap [t,\infty)$. Weak invariance is called semi-invariance in [2] and [3].

The equations (E) and (C) are closely related and many results that are true for (E) can be extended to (C). (C) is more general since we allow $\Omega(t,x) = \{f(t,x)\}$. Sometimes it is useful to study (C) to get results for (E). We shall not specify which equation we are studying since these results hold for both equations.

*) This work was supported in part by the National Science Foundation, NSF Grant GP 6114.

Definition. We say $Q \in R \times R^d$ is locally compact if each $(t,x) \in Q$, has a neighborhood $N_{t,x}$ such that $N_{t,x} \cap Q$ is compact. An equivalent condition is that there exists an open set Q^o and a closed set Q^c such that $Q^o \cap Q^c = Q$. In particular since U is open, if Q is closed in U, i.e. $\overline{Q} \cap U = Q$, then Q is locally compact.

We say $v \in R^d$ is subtangential to Q at $(t,x) \in Q$ if

$$\liminf_{s \to 0^+} d((t+s,\ x+sv),Q) = 0$$

where $d(.,.)$ denotes the Euclidean distance in R^{d+1}, and if $z = (t,x)$, then $d(z,Q) = \inf_{q \in Q} d(z,q)$. In particular, if (t,x) is in the interior of Q, then any v is subtangential to Q at (t,x). If $(t,x) \in \partial Q$, then intuitively v is subtangential to Q at (t,x) if $(1,v)$ is either "tangent" to Q at (t,x) or points into Q. Actually, being tangent is not defined since no smoothness assumption on Q has been made. We say Ω is weakly subtangential to $Q \subset U$ if for each $(t,x) \in Q$ there exists $v \in \Omega(t,x)$ such that v is subtangential to Q at (t,x).

Theorem. Let $Q \subset U$ be locally compact and let Ω be compact upper semi-continuous.

(i) If Ω is not weakly subtangential to Q, then Q is not weakly invariant.

(ii) If (t,x) is convex for each $(t,x) \in Q$ and Ω is weakly subtangential for Q, then Q is weakly invariant.

The above theorem for the case of ordinary differential equations, i.e. $\Omega(t,x) = \{f(t,x)\}$ was proved by Nagumo in a little known paper [4]. The theorem has many applications for ordinary differential equations (See [1]) and for contingent equations though Nagumo did not apply his theorem. Weakly invariant sets have been frequently discussed, though primarily for dynamical systems and autonomous differential equations with uniqueness of solutions, in which case weak invariance is equivalent to invariance (i.e., all solutions through a set Q remain in Q). Cellina [5] has applied this Theorem to the problem of dealing with branching external solutions in control equations.

BIBLIOGRAPHY

[1] JAMES A. YORKE, Invariance for ordinary differential equations, submitted to Mathematical Systems Theory.

[2] AARON STRAUSS and JAMES A. YORKE, On Asymptotically autonomous differential equations, Mathematical Systems Theory, 1(1967) 175-182.

[3] TARO YOSHIZAWA, Stability theory by Liapunov's second method, Math. Soc. of Japan, Tokyo, 1966.

[4] M. NAGUMO, Uber die Lage der Integralkurven gewöhnlicher Differentialgleichungen, Proc. Phys.-Math. Soc. Japan 24(1942) 551-559.

[5] ARRIGO CELLINA, A note on the Pontriagin maximum principle, to be published.

SPACES OF SOLUTIONS *)

James A. Yorke

INTRODUCTION

We begin by studying the non-autonomous ordinary differential equation

$$\text{(E)} \qquad \dot{x} = f(t,x)$$

where $U \subset R \times R^d$ is open, and $(t,x) \in U$, and $f : U \longrightarrow R^d$ is continuous. For a curve ϕ we let D_ϕ denote its domain, which is an interval. By a solution ϕ we always will mean a "non-continuable" solution; i.e., there is no solution Ψ (defined on any interval D_Ψ) such that $D_\phi \subset D_\Psi$, $D_\Psi \neq D_\phi$, and $\Psi \equiv \phi$ on D_ϕ. We are also interested in studying other systems with non-uniqueness including the contingent equation $\dot{x} \in \Omega(t,x)$, and the control equation $\dot{x} = g(t,x,u)$ for $u \in \Lambda$, where various additional conditions may be put on the integrable function $u(\cdot)$ or on the set $g(t,x,\Lambda)$. In general we are interested in the set S of curves or solutions (with varying domains) which represent some kind of non-autonomous dynamical system without uniqueness.

Hartman proves the following convergence theorem for ordinary differential equations. See [1, Chapter II, Theorem 3.2]. A version was proved by Kamke [14]. This Convergence Theorem is stated more easily if we introduce the following concept of compact convergence.

<u>Definition.</u> Let $\{\phi_n\}$ be a sequence of curves with interval domains $D_{\phi_n} \subset R$ and let ϕ be a curve with domain D_ϕ. We say ϕ_n <u>converges compactly to</u> ϕ, $\phi_n \xrightarrow{c} \phi$, if for each compact set $K \subset D_\phi$, all but finitely many D_{ϕ_n} contain K and $\phi_n(t) \longrightarrow \phi(t)$ as $n \longrightarrow \infty$ uniformly for $t \in K$. Write $(t_n,\phi_n) \xrightarrow{c} (t,\phi)$ if $t_n \longrightarrow t$ and $\phi_n \xrightarrow{c} \phi$.

*) This work was supported in part by the National Science Foundation under Grant NSF GP 6114.

Convergence Theorem. Let $\{\phi_i\}$ be a sequence of solutions of (E) and let $t_i \in D_{\phi_i}$ be such that $(t_i, \phi_i(t_i)) \longrightarrow (t,x)$ as $i \to \infty$ for some $(t,x) \in U$. Then there is a solution ϕ of (E) such that $\phi(t) = x$ and there is a subsequence $\{\phi_{i_n}\}$ such that $\phi_{i_n} \xrightarrow{c} \phi$.

Let us point out that since the above D_ϕ contains some interval $[t-\varepsilon, t+\varepsilon]$ for some $\varepsilon > 0$, it follows that all but finitely many of the ϕ_{i_n} are defined on $[t-\varepsilon, t+\varepsilon]$. We might further say that "in the limit" $D_\phi \subset \lim D_{\phi_{i_n}}$. Although we might have $D_{\phi_i} = R$ for all i we may also have $D_\phi = (-\infty, 0)$; (see Example 1.2). If the solution ϕ of (E) such that $\phi(t) = x$ is unique, it follows that $\phi_i \xrightarrow{c} \phi$. Note that if ϕ_n and ϕ are the real-valued curves given by $\phi_n(t) = n^{-1}e^t$ and $\phi(t) \equiv 0$, then $\phi_n(t) \to \phi(t) = 0$ uniformly for t in each compact $K \subset R$, but not uniformly for $t \in R$; hence we have precisely $\phi_n \xrightarrow{c} \phi$.

The Convergence Theorem is very basic in that many theorems for an autonomous or non-autonomous equation (E) can be derived from it. Furthermore, it holds true for many other types of systems quite different from (E).

We say ϕ is a <u>curve</u> if D_ϕ is an interval and ϕ is a continuous function $\phi : D_\phi \longrightarrow X$ where X is some fixed topological space. Let S be a family of curves and let $W \subset R \times X$ be such that $(t, \phi(t)) \in W$ for all $\phi \in S$ and $t \in D_\phi$. We then say S is a <u>family of curves on</u> W. The concept of convergence $\phi_n \xrightarrow{c} \phi$ actually defines a topology on S. For $Q \subset W$, we write $S*Q = \{(t,\phi) : \phi \in S,\ (t,\phi(t)) \in Q, \text{ and } t \in \text{interior } D_\phi\}$. Then S*W is our main topological space. We define in Section 2 the pair (S,W) to be a <u>curve space</u> provided that one axiom is satisfied, and that axiom is essentially the Convergence Theorem. For (E) we have $X = R^d$ and $W = U$.

We study curve spaces from two points of view. In Section 1 we study (E) and show how two important thorems of ordinary differential equations can be restated quite simply using this topology, and note how their statements become very similar in the terminology of curve spaces. The topological space of solutions with initial conditions S*U offers a new tool for the study of ordinary differential equations. In Section 4 we show how the space of solutions allows us to get new theorems about differential equations, that is, new theorems which can be stated in classical terms.

We solve a problem of C.C. Pugh on solutions funnel cross-sections (Theorem 4.3). It is hoped that the theorems given in Section 4 for an ordinary diffential equation state completely all the topological properties of a differential equation which do not depend upon differentiability and the metric of U, just as the axioms of a dynamical system seem to give all the topological properties for autonomous differential equations with uniqueness when $D_\phi = R$ for all solutions ϕ .

Our second point of view is the study of an abstract curve space as an abstract axiomatic system representing many systems without uniqueness. In Section 2 we examine the basic curve space with no additional axioms. Theorems 2.1 and 2.2 and Corollary 2.1 show that each curve of a curve space has the same kind of geometric behavior as a solution of a differential equation. Curve spaces represent a broad family of systems that in general do not have uniqueness , and may be non-autonomous, and need not have solutions defined on the whole real line, as is shown by Examples 2.1 through 2.5.

We also define in Section 2 the concept of autonomous curve space (ACS). Essentially a curve space is autonomous if for each $\phi \in S$ each translate ϕ_t is in S, where $\phi_t(t+s) = \phi(s)$ for $s \in D_\phi$. Many of the properties for dynamical systems and autonomous ordinary differential equations also hold for an ACS. We define "positive limit set", "x_0 is a rest point", "invariant" and "weakly invariant" sets (a set Q_0 in the range space X is weakly invariant, if for each $x \in Q_0$ there is a $\phi \in S$ such that $\phi(0) = x$ and $\phi(t) \in Q_0$ for all $t \in D_\phi$), "minimal set" (minimally weakly invariant set), and "recurrent function".

Many of the proofs for dynamical systems are almost the same as for an ACS. It is quite surprising how much of the theory can be made independent of the uniqueness assumption of dynamical systems and follows purely from the convergence properties of an ACS. We are of course limited in the size of this paper and so we only sketch the main features of dynamical systems which are extended. We make no attempt here to develop the theory of an ACS other than to generalize dynamical systems and ordinary differential equations. We mention:

The limit set of a curve in an ACS is weakly invariant and if it is compact, it is connected (Theorem 2.5); if the limit set equals the trajectory, the curve must repeat itself in a way that for dynamical systems would be periodic (Theorem 2.6). The closure of an invariant set is weakly invariant (Theorem 2.7). Any curve remaining in a minimal set must be recurrent (Theorem 2.9), and if ϕ is recurrent, then $\overline{\phi(D_\phi)}$ is a compact weakly invariant set (Theorem 2.10). One gets the usual theorem for dynamical systems that at least one minimal set exists in each compact weakly invariant set (Theorem 2.8).

In Section 3 we return to non-autonomous systems and show that additional results are true for curve spaces when we assume one or both of a new pair of axioms: we show that Kneser's and Fukuhara's theorems on solution funnels hold (Theorem 3.1 and 3.2). We state a necessary and sufficient condition for an ACS to have a constant curve $\phi(t) \equiv x_o$ for a given point x_o, i.e. for x_o to be a rest point (Theorem 3.2). The boundary of an invariant set is weakly invariant (Theorem 3.5 ii). An invariant minimal set with interior must be an entire component of X (Theorem 3.5 iii); the boundary of the image of a set is in the image of the boundary (Theorem 3.5 iv). The Poincaré-Bendixson Theorem is generalized in Theorems 3.7 and 3.8. Two conjectures are made, in Sections 3 and 4.

The results stated in this paper will be proved elsewhere.

1. THE SPACE OF SOLUTIONS OF A DIFFERENTIAL EQUATION

Let U be an open subset of $R \times R^d$. We consider the equations

$$\text{(E)} \qquad \dot{x} = f(t,x)$$

$$\text{(C)} \qquad \dot{x} \in \Omega(t,x)$$

where $f : U \longrightarrow R^d$ is continuous and where $\Omega(t,x)$ is a compact convex non-empty subset of R^d and Ω is an upper-semi-continuous set function i.e., $\{(v,t,x) : v \in \Omega(t,x)$ and $(t,x) \in U\}$ is a compact subset of $R^d \times U$ when U is compact. By a solution of (C) we mean an absolutely continuous function ϕ defined on an interval D_ϕ such that $\dot{\phi}(t) \in \Omega(t,\phi(t))$ for almost all $t \in D_\phi$. By "solution" of (E) or (C) we always mean <u>non-continuable</u> solution (as before).

We shall not specify which of the equations we are considering except when there is a special reason to make a distinction. S will denote the set of all solutions. In this section Q will denote a subset of U, and $U \subset R \times R^d$ corresponds to $W \subset R \times X$. We will write S_Q for $\{\phi \in S :$ for some t, $(t,\phi) \in S^*Q\}$. S^*U and $S^*Q \subset S^*U$ are topological spaces, with the topology given by $\xrightarrow{c}$.

<u>Example 1.1.</u> If all curves in S have domain R, then the topology on S is the topology of uniform convergence on compact sets, also called the compact-open topology. As mentioned in the introduction we cannot expect uniform convergence on all of R as the following differential equation shows: $\dot{x} = x$ has solutions ae^t for $a \in R$. Note that if $a_n \longrightarrow 0$, then $a_n e^t \longrightarrow 0$ as $n \longrightarrow \infty$ uniformly for t in each compact $K \subset R$, but not uniformly on R.

<u>Exmaple 1.2</u>. When the domains D_ϕ of the curves ϕ vary, then the topology on S offers problems. Consider the equation $\dot{x} = -2tx^2$, which has solutions $\phi_n(t) = (t^2 + n^{-1})^{-1}$ for $n = 1,2,\ldots$, with $D_{\phi_n} = R$, and also $\phi_+(t) = t^{-2}$ with $D_{\phi_+} = (0,\infty)$ and $\phi_-(t) = t^{-2}$ with $D_{\phi_-} = (-\infty, 0)$. Since $\phi_+(t) \longrightarrow \infty$ and $\phi_-(t) \longrightarrow \infty$ as $t \longrightarrow 0$, ϕ_+ and ϕ_- cannot be defined on larger domains. If $K \subset (0,\infty)$ is compact, then $\phi_n(t) \longrightarrow \phi_+(t)$ uniformly on K so $\phi_n \xrightarrow{c} \phi_+$, and similarly $\phi_n \xrightarrow{c} \phi_-$. Hence $\{\phi_n\}$

converges to two distinct solutions and the topology on S is not Hausdorff. Note however that if $\{t_n\} \subset R$ and $t_+ > 0$ are chosen so that $(t_n,\phi_n) \xrightarrow{c} (t_+,\phi_+)$, then $t_+ \notin D_{\phi_-}$ and $(t_+,\phi_-) \notin S^*(R\times R)$. The topology on S*U is in fact Hausdorff.

Example 1.3. If solutions are unique, then for each $(t,x) \in U$, there exists a unique $\phi_{t,x}$ such that $\phi(t) = x$, and the function $h : S^*W \longrightarrow W$, given by $h(t,\phi) = (t,\phi(t))$ is a homeomorphism.

Theorem 1.1. *If* Q *is compact, then* S*Q *is compact.*

Theorem 1.2. *If* Q *is connected, then* S*Q *is connected.*

Theorem 1.1 is a restatement of the Convergence Theorem since it can be shown that S*U is a metric space.

Theorem 1.2 is a generalization of Kneser's Theorem which says (using our symbols) if Q is connected and T is such that for all $\phi \in S_Q$, $\phi(T)$ is defined, then $J = \{\phi(T) : \phi \in S_Q\}$ is connected. Kneser's Theorem follows from Theorem 1.2 since the function $e_T : S^*Q \longrightarrow J : (t,\phi) \longrightarrow \phi(T)$ is continuous. The continuous image J of a connected set S*Q is connected. Hukuhara (Fukuhara) [2] proved a theorem quite similar to Theorem 1.2, in which he dealt with a particular contingent equation and assumed Q was also compact and that all solutions of his equation were defined on a compact interval I. By restricting attention to I, he considered the space of solutions using the uniform topology (sup norm). We now state a corollary of Theorem 1.2 to indicate how it contains information not contained in Kneser's Theorem.

Corollary 1.1 *Let* Q *be connected and assume* S_Q *contains more than one solution.*

(i) *Choose* $\phi \in S_Q$ *and let* K *be a compact subset of* D_ϕ. *Then for any* $\epsilon > 0$ *there exists* $\Psi \in S_Q$ *distinct from* ϕ *such that* $K \subset D_\Psi$ *and*

$$|\Psi(t) - \phi(t)| < \epsilon \quad \text{for all} \quad t \in K .$$

(ii) *Let* $S_Q = S_1 \cup S_2$ *where* S_1 *and* S_2 *are non-empty subsets of* S *and* assume $K \subset R$ is a compact set such that $K \subset D_\phi$ for all $\phi \in S_Q$. Then for any $\epsilon > 0$ there exists $\phi_1 \in S_1$ and $\phi_2 \in S_2$ such that

$$\sup_{t \in K} |\phi_1(t) - \phi_2(t)| < \epsilon .$$

At this point we describe one more result which can be obtained for (C) if an additional condition is satisfied.

Definition 1.2. A set Y is contractible (through itself to a point) if there exists a continuous function $h : [0,1] \times Y \longrightarrow Y$ such that for all $y \in Y$, $h(0,y) = y$ and $h(1,y) \equiv \eta$ for some $\eta \in Y$.

Note that a contractible set is always connected, arc-wise connected and simply-connected.

Theorem 1.3. Assume there is a C^1 function $g : U \longrightarrow R^d$ such that for all $(t,x) \in U$, $g(t,x) \in \Omega(t,x)$. Then if Q is contractible, S^*Q is contractible.

There are examples in which no such g exists, with $\Omega(t,x) = \{f(t,x)\}$ and $Q = \{(t_0,x_0)\}$, and $S^*(t_0,x_0)$ is not arc-wise connected, let alone contractible. This example is by Pugh [4]. (See Section 4). Actually he gives an example in which the reachable set J is not arc-wise connected; hence S^*Q cannot be arc-wise connected in his example since J is the continuous image of $S^*(t_0,x_0)$. $S^*(t_0,x_0)$ is a very nice topological set except with respect to "arc-wise properties". In fact $S^*(t_0,x_0)$ is "almost contractible" or as we define in Section 4, it is co-contractible.

2. ABSTRACT CURVE SPACES

For the remainder of this paper, W will denote an arbitrary subset of $R \times X$, where X is a metric space. Q will denote a subset of W. Let S be a family of curves in W. Again we write $\phi \in S_Q$ if for some t, $(t,\phi) \in S^*Q$.

The topology on S^*W is the topology given in terms of the convergence $\xrightarrow{c}$ defined in the introduction. An alternate way to define the topology on S is as follows. For each $\epsilon > 0$ and $\phi \in S$ and compact $K \subset D_\phi$, the following set is a neighborhood of ϕ.

$$\phi(\epsilon,K) = \{\Psi \in S : K \subset D_\Psi \text{ and } |\phi(t) - \Psi(t)| < \epsilon \text{ for all } t \in K\}$$

For each ϕ, the class of all $\phi(\epsilon,K)$ such that $\epsilon > 0$ and $K \subset D_\phi$ is a base for the neighborhood system of ϕ. The topology generated by $\{\phi(\epsilon,K)\}$ is the weakest topology (i.e. fewest open sets) such that ϕ_n converges to ϕ in this topology iff $\phi_n \xrightarrow{c} \phi$.

Hajek [9] has developed a theory of "processes" and non-topological "solution spaces"; Roxin has given non-unique dynamical systems [6]; see Sell [7] for non-autonomous systems and other references in [5, p. 297].

<u>Compactness Axiom.</u> For all compact $Q \subset W$, S^*Q is compact.

<u>Definition 2.1.</u> The pair (S,W) is a <u>curve space</u> if S is a family of curves in W and the Compactness Axiom is satisfied. For the remainder of this paper (S,W) will be a curve space. We say S <u>has existence on</u> Q if for each $(t,x) \in Q$, there is some $\phi \in S$ with $\phi(t) = x$. S <u>has uniqueness</u> on Q if for each $(t,x) \in Q$, there is at most one such $\phi \in S$. We say S <u>has domain</u> R if for all $\phi \in S$, $D_\phi = R$. Note that in general S might be empty.

<u>Examples.</u> (S,W) is a curve space if S is the set of all

(2.1) solutions of (E) or (C) where $W = U$.

(2.2) solutions of (E) such that $(s,\phi(s)) \in W$ for all $s \in D_\phi$ where $W \subset U$ is assumed relatively closed in U.

(2.3) solutions ϕ satisfying (G), $\frac{dx}{ds} = g(s,x,u)$, for some constant $u = u_\phi$ where $(s,x) \in W$ and $W \subset R \times R^d$ is open and u is in a compact set A and $g : W \times A \longrightarrow R^d$ is continuous. A can be a finite set.

(2.4) solutions of (G) where g,W, and A are as in (2.3) except that $u_\phi(\cdot)$ is a function of $s \in D_\phi$ and satisfies one of the additional condition which follow: (a) There is a fixed constant L and for each ϕ, u_ϕ is Lipschitz continuous with constant L; or we may assume all u_ϕ are piecewise constant such that (b) there is a fixed upper bound N on the number of discontinuities or (c) the discontinuities can occur only on a closed discrete set $\{s_i\} \subset R$, or (d) the distance between discontinuities of u_ϕ must be larger than some given $\varepsilon > 0$. We cannot however allow u_ϕ to be any measurable function unless we also assume $g(s,x,\Lambda)$ is convex for $(s,x) \in W$. In general the class $\{u(\cdot)\}$ will be compact in the L^1 convergence topology.

(2.5) the curves ϕ_x of a dynamical system; that is, if $\pi : R \times X \longrightarrow X$ is a dynamical system, X a topological space, then for each x, we have a curve ϕ_x given by $\phi_x(s) = \pi(s,x)$ for all $s \in R = D_{\phi_x}$. Here $W = R \times X$.

Example 2.4 represents control equations in which the control $u(\cdot)$ is restricted in common ways. One of the most frequent complaints about control theory is that the assumption that u need be only measurable is not physically realizable. When u is measurable, (G) becomes a contingent equation with $\Omega(t,x) = g(t,x,\Lambda)$, and in order to satisfy the Compactness Axiom in general, we then must assume $g(t,x,\Lambda)$ is convex". The solutions of certain functional differential equations, as $\dot{x}(t) = f(x(t), x(t-1))$, can be adapted to our scheme.

Theorem 2.1 and 2.2 and Corollary 2.1 show that the curves in S have several of the basic properties that hold for solutions of (E). Lemma 2.1 and 2.2 give useful tools for proofs. Lemma 2.2 says that the evaluation maps are continuous when they are defined.

<u>Theorem 2.1.</u> <u>Let</u> $\phi \in S$. <u>Then</u> D_ϕ <u>is open, and</u> $(t,\phi(t))$ <u>has no limit points in</u> W <u>as</u> $t \longrightarrow \sup D_\phi$ <u>and as</u> $t \longrightarrow \inf D_\phi$.

<u>Corollary 2.1.</u> <u>Let</u> $\phi \in S$ <u>and</u> $W = R \times R^d$. <u>If</u> $\sup D_\phi = \omega < \infty$, <u>then</u> $|\phi(t)| \rightarrow \infty$ <u>as</u> $t \longrightarrow \omega$.

<u>Theorem 2.2.</u> <u>For any compact</u> $Q \subset W$ <u>there exists an</u> $\epsilon_Q > 0$ <u>such that for all</u> $(t,\phi) \in S^*Q$, $[t-\epsilon_Q, t+\epsilon_Q] \subset D_\phi$.

<u>Lemma 2.1.</u> If W is a locally compact metric space, then S*W is a locally compact metric space.

<u>Lemma 2.2.</u> Let $\beta : Q \longrightarrow R$ be continuous. Then the functions defined as follows are continuous,

$$E_1 : S^*W \longrightarrow X \text{ given by } E_1(t,\phi) = \phi(t) ,$$

$$E_2 : S^*W \longrightarrow W \text{ given by } E_2(t,\phi) = (t,\phi(t)) ,$$

$$E_3 : S^*Q \longrightarrow X \text{ given by } E_3(t,\phi) = \phi(\beta(t,\phi(t))) .$$

A dynamical system (Example 2.5) is an "autonomous curve space". The theory of autonomous systems is much richer than that for non-autonomous systems because we can study many "natural objects" such as limit sets, constant solutions, minimal sets. First we define an autonomous curve space and point out that if strong restrictions are made, it reduces to a dynamical system.

<u>Definition 2.2.</u> The pair (A,X) is an <u>autonomous curve space</u> (ACS) if

(i) $(A, R \times X)$ is a curve space,

(ii) for any curve $\phi \in A$ and $\tau \in R$, ϕ_τ is in A where ϕ_τ is defined by $\phi_\tau(s) = \phi(s+\tau)$ for all $s + \tau \in D_\phi$ and $D_\phi = \tau + D_{\phi_\tau}$ (translation by τ).

For the remainder of this paper (A,X) will be an ACS.

<u>Theorem 2.3.</u> <u>An autonomous curve space having existence, uniqueness, and domain</u> R <u>is equivalent to a dynamical system; that is, if we define</u> $\pi(t,x)$ <u>equal to</u> $\phi_x(t)$ <u>where</u> $\phi_x \in A$ <u>is chosen such that</u> $\phi_x(0) = x$, <u>then</u> $\pi : R \times X \longrightarrow X$ <u>is continuous</u>, $\pi(0,x) = x$, <u>and</u> $\pi(t_1,\pi(t_2,x)) = \pi(t_1+t_2,x)$ <u>for all</u> t_1 <u>and</u> t_2 <u>in</u> R <u>and</u> $x \in X$.

We say $x_0 \in X$ is a <u>rest point</u> for (A,X) if there is some $\phi \in A$ such that $\phi(t) \equiv x_0$ for $t \in D_\phi$. By Theorem 2.1 we in fact have $D_\phi = R$. The following theorem is not well known. It was first proved by Ura and Kimura [10, Theorem 3] for dynamical systems and does not seem to have appeared anywhere for systems without uniqueness. Theorem 3.3 gives an improved version.

Theorem 2.4. Let X be locally compact and let $x_0 \in X$ be a rest point such that x_0 is not an isolated point of X. Assume A has existence on X; then either

(i) there is a non-constant $\Psi \in A$ such that $\Psi(0) = x_0$ or,

(ii) for any neighborhood N of x_0 there is a $\Psi_N \in A$ with D_{Ψ_N} containing either $(-\infty,0]$ or $[0,\infty)$ such that $\Psi_N(t)$ is in $N - \{0\}$ on that interval.

We now show that many more of the results which seem to be a fundamental part of the theory for a dynamical system, as presented by Nemytskii and Stepanov [11] or Bhatia and Szegö [5], in fact also hold for an ACS. Our definition of the positive limit set of a curve is standard. Invariance can be defined two ways for systems without uniqueness. Minimal set here is defined using weak invariance because this definition seems to give results closest to those for dynamical systems. When A has uniqueness, then our definitions reduce to the usual ones. We will return to some of these concepts in Section 3 with two additional axioms.

Definition 2.3. For $\phi \in A$, the positive (or omega) limit set of ϕ, $\Lambda^+(\phi)$, is the set of $y \in X$ such that there exists $\{t_n\} \subset D_\phi$, $t_n \longrightarrow \infty$, such that $\phi(t_n) \longrightarrow y$ as $n \longrightarrow \infty$. The trajectory of ϕ will be written $\phi(D_\phi) = \{\phi(s) : s \in D_\phi\}$. A set $W_0 \subset X$ is weakly invariant (or semi-invariant) if for each $x \in W_0$ there exists $\phi \in A$ such that $x \in \phi(D_\phi) \subset W_0$, and $I \subset X$ is invariant if for each $\phi \in A$ either $\phi(D_\phi) \cap I$ is empty or $\phi(D_\phi) \subset I$. A set M is minimal if M is non-empty, closed, weakly invariant, and no proper subset is non-empty, closed and weakly invariant. Let $B \subset A$. Define $\Lambda^+(B)$ to be the set of $y \in X$ such that there exist $\phi_n \in B$ and $t_n \in D_{\phi_n}$ such that $t_n \longrightarrow \infty$ and $\phi_n(t_n) \longrightarrow y$.

Note that $\Lambda^+(B) \supset \bigcup_{\phi \in B} \Lambda^+(\phi)$ but these sets are not always equal. Note that $\Lambda^+(\phi) = \Lambda^+(\{\phi\})$. The limit set of a family of solutions was defined by Strauss and Yorke [12]. See [12] for applications. The limit set of a point x_0 may be defined as the limit set of the family of curves ϕ such that $\phi(0) = x_o$, or alternatively as the union the limit sets of all such curves. Note that if I is invariant, it is weakly invariant if and only if A has existence on I.

Theorem 2.5. For any $B \subset A$, $\Lambda^+(B)$ is weakly invariant. If X is locally compact and $\Lambda^+(\phi)$ is compact, then $\Lambda^+(\phi)$ is connected.

Theorem 2.6. Let X be locally compact or be a complete metric space. If for some $\phi \in A$, $\Lambda^+(\phi) = \phi(D_\phi)$, there exists $t_1 \in D_\phi$, $t_2 \in D_\phi$, $t_1 \neq t_2$, such that $\phi(t_1) = \phi(t_2)$.

Theorem 2.7. If $W \subset X$ is weakly invariant, then $\overline{W}$ is weakly invariant.

A minimal M has the property that if $\phi \in A$ and $\phi(D_\phi) \subset M$, then $\overline{\phi(D_\phi)} = M$.

Theorem 2.8. If $W_o \subset X$ is a non-empty compact weakly invariant set, then there is a minimal set $M \subset W_o$.

Definition 2.4. Let X be a metric space with metric $d(\cdot,\cdot)$. A curve ϕ is recurrent if for each $t \in D_\phi$ and $\varepsilon > 0$ there exists $T = T(t,\varepsilon) > 0$ such that for all $s \in D_\phi$, $d(\phi(t),\phi([s-T,s+T])) < \varepsilon$.

Theorem 2.9. Let $M \subset X$ be a compact minimal set. Then for each $\phi \in A$ such that $\phi(D_\phi) \subset M$, $\Lambda^+(\phi) = M$, and if X is a metric space, then ϕ is recurrent.

Theorem 2.10. Let X be a complete metric space. If $\phi \in A$ is recurrent, then $\overline{\phi(D_\phi)}$ is compact and weakly invariant.

In Theorem 2.10, $\overline{\phi(D_\phi)}$ need not be a minimal set, but it is a minimal set for (B,X) for some $B \subset A$ such that (B,X) is an ACS.

It is natural to ask if various stability results can be proved for rest points of an ACS. Theorem 2.11 and 3.4 give two results concerning stable attractors and stable rest points for X a metric space.

Definition 2.5. A point x_0 is stable (in the sense of Liapunov) if for any $\epsilon > 0$ there exists a $\delta > 0$ such that for $\phi \in A$ with $d(\phi(t), x_0) < \delta$ for some $t \in D_\phi$, $d(\phi(\tau), x_0) < \epsilon$ for all $\tau \geq t$, $\tau \in D_\phi$. The point x_0 is an attractor if for some compact neighborhood N of x_0, whenever $\phi(t) \in N$ for $\phi \in A$ and $t \in D_\phi$, we have $\Lambda^+(\phi) = \{x_0\}$. We say x_0 is a uniform attractor if x_0 is an attractor and X is a metric space and for all $\delta > 0$, there exists $T(\delta)$ such that if $(0, \phi) \in A^*N$, then $[0, \infty) \subset D_\phi$ and $d(\phi(t), x_0) < \delta$ for $t > T(\delta)$.

Theorem 2.11. Let X be a locally compact metric space and let x_0 be a stable attractor. Then x_0 is a uniform attractor.

3. TWO AXIOMS

In Section 2 we found that a curve space (particularly an ACS) satisfies many properties of differential equations and dynamical systems. We now introduce two additional axioms. The Switching Axiom is not satisfied by Examples 2.3 and 2.4, yet it seems very basic for an ordinary differential equation, and it seems rather surprising that it was not needed for Section 2. The Connectedness Axiom is satisfied by all five examples, provided that in 2.3 and 2.4 Λ is connected.

Switching Axiom (SA). For curves ϕ_1 and ϕ_2 and $t \in D_{\phi_1} \cap D_{\phi_2}$ such that $\phi_1(t) = \phi_2(t)$, define $\Psi(s) = \phi_1(s)$ for $s \leq t$ and $\Psi(s) = \phi_2(s)$ for $s \geq t$ (with D_Ψ defined appropriately). If $\phi_1 \in S$ and $\phi_2 \in S$, then $\Psi \in S$.

Connectedness Axiom (CA). If Q is connected, then S*Q is connected.

The best known topological theorems for differential equations other than Theorems 2.1 and 2.2 and Corollary 2.1 are the theorems of Kneser and Fukuhara, which are stated here as Theorems 3.1 and 3.2. We cannot continue to dwell on non-autonomous systems because there is little to deal with. We could develop Liapunov stability theory and show the existence of Liapunov functions but we prefer not to take that direction except for Theorem 3.4 iii for an ACS. We thus return to our study of an ACS. For every theorem in which we assume CA or SA or whenever anywhere in this paper we assume that W or X is locally compact, counterexamples can be shown if the condition is dropped.

Theorem 3.1. Let (S,W) satisfy CA. Let Q be connected and assume that for some T and all $\phi \in S_Q$, $T \in D_\phi$. Then the "T-reachable set for Q" $J = \{\phi(T) : \phi \in S_Q\}$ is connected. If Q is compact, then J is compact.

For point $(t,x) \in W$, we define the solution funnel $F_{t,x} = \{(\tau,\phi(\tau)) : \phi \in S\{(t,x) \text{ and } \tau \in D_\phi\}$. We denote the boundary of a set by ∂. The following theorem was proved by Fukuhara for ordinary differential equations. Theorem 3.5 iv is closely related.

Theorem 3.2. *Let* (S,W) *satisfy* CA *and* SA. *Let* W *be locally compact. Let* $(t,x) \in W$ *and let* $(t_1,x_1) \in \partial F_{t,x}$ *with* $t_1 > t$ *such that if* $\Psi(t) = x$, *then* $t_1 \in D_\Psi$. *Then there exists* $\phi \in S$ *such that* $\phi(t) = x$, $\phi(t_1)$ x_1, *and* $(\tau,\phi(\tau)) \in \partial F_{t,x}$ *for all* $\tau \in [t,t_1]$.

Theorem 3.3. *Let* X *be locally compact and let* (A,X) *be an* ACS *satisfying* SA. *Then* $x_0 \in X$ *is a rest point if and only if* x_0 *is either an isolated point or for every neighborhood* N *of* x_0 *there is a curve* Ψ *defined on and remaining in* N *on* $[0,\infty)$ *or* $(-\infty,0]$, *where* Ψ *is not identically equal to* x_0.

Theorem 3.4. *Let* (A,X) *be an* ACS *satisfying* SA. *Let* X *be locally compact. The following conditions are equivalent:*

(i) $x_0 \in X$ *is stable.*

(ii) $\{x_0\}$ *is the intersection of positively invariant neighborhoods of* x_0.

(iii) *there exists a function* $V : N \longrightarrow [0,\infty)$ *for some neighborhood* N *of* x_0 *satisfying* (a) $V(x) = 0$ *iff* $x = x_0$, (b) *for any sequence* $\{x_n\}$, $V(x_n) \longrightarrow 0$ *iff* $x_n \longrightarrow x_0$ *as* $n \longrightarrow \infty$, *and* (c) *for each* $\phi \in A$ *such that* $\phi(t_1) \in N$ *and* $\phi(t_2) \in N$ *and* $t_2 > t_1$, $V(\phi(t_2)) \leq V(\phi(t_1))$.

Theorem 3.5. *Let* (A,X) *satisfy* SA *and* CA.

(i) *Let* W_1 *and* W_2 *be closed and weakly invariant such that* $W_1 \cup W_2$ *is invariant. Then* $W_1 \cap W_2$ *is weakly invariant.*

(ii) *If* $I \subset X$ *is an invariant set with* existence then ∂I *is weakly invariant.*

(iii) *If* M *is an invariant minimal set and* M *has an interior point, then* M *is an open set, and if* X *is connected, then* M = X.

(iv) *Let* $X_0 \subset X$ *be compact, and assume that if* $\phi \in A$ *and* $\phi(0) \in X_0$, *then* $\phi(T)$ *is defined. Write for* $Y \subset X$, $A(T,Y) = \{\phi(T) : \phi \in A$ *and* $\phi(0) \in Y\}$. *Then* $A(T,\partial X_0)$ $\partial A(T,X_0)$.

The result (iii) was proved for dynamical systems by G.Ts. Tumarkin. It follows from (iii) that if the assumptions of 3.4 iii are satisfied and X is R^d, then M = X. The formulation for dynamical systems in [11] is given for *compact* minimal sets but the proof does not require compactness.

Theorem 3.6. Let (A,X) satisfy SA and CA and let X be locally compact. Let $Q_0 \subset X$ be connected and let $B = \{\phi \in A : \phi(0) \in Q_0\}$. Then if $\Lambda^+(B)$ is compact, it is connected. If $\Lambda^+(B)$ is not compact, then no connected component of $\Lambda^+(B)$ is compact.

Poincaré-Bendixson theorem is historically one of the earliest results (and is still one of the nicest) of the qualitative theory of ordinary differential equations so it is only fair to point out what kind of a generalization we can get. Of course a certain amount of the usual theory depends on uniqueness so we cannot get the usual theorems; however, when the additional assumption of uniqueness is made, our theorems almost immediately reduce to the usual statements.

Theorem 3.7. Let (A,X) satisfy SA and let X be an open subset of R^2. Let $\phi \in A$ be chosen such that $\Lambda^+(\phi)$ is compact. Suppose $\Lambda^+(\phi)$ contains no rest points. Then there is a family $P \subset A$ of periodic curves such that

$$\bigcup_{\Psi \in P} \Psi(D_\Psi) = \Lambda^+(\phi) .$$

Theorem 3.8. Let (A,X) satisfy SA and let $X \subset R^2$ be open and simply connected. Let $\Lambda^+(\phi)$ be compact and non-empty for some $\phi \in A$. Then X has a rest point.

To finish this section we include a conjecture. We say that a set $N \subset X$ has a section if for some $B \subset N$ there is a $\tau > 0$ such that for every ϕ with $\phi(0) \in N$, there is a unique $T(\phi) \in D_\phi$ such that $|T(\phi)| < \tau$ and $\phi(T(\phi)) \in B$ and the function T is continuous where it is defined. For a dynamical system every point that is not a rest point has a neighborhood with a section.

Conjecture. Let (A,X) satisfy SA and let X be locally compact. If $x \in X$ is not a rest point, then x has a neighborhood N_x with a section.

4. SOLUTION FUNNELS OF DIFFERENTIAL EQUATIONS

Let S denote the family of solutions of (E) or (C). Fix some $(t_o,x_o) \in U$. Define for $Q \subset U$ the <u>funnel for</u> Q to be

$$F(Q) = \{(\tau,\phi(\tau)) : \tau \in D_\phi,\ \phi \in S_Q\}, \quad \text{and}$$

$$F_o = F(\{(t_o,x_o)\}) \ .$$

Fix some $t_1 > t_o$ such that $\phi(t_1)$ is defined for all ϕ such that $\phi(t_o) = x_o$. Define the <u>F-section for</u> (t_o,x_o) (solution funnel cross-section at time t_1) to be

$$J = \{\phi(t_1) : \phi(t_o) = x_o\} \ .$$

A set $J_1 \subset R^d$ is an <u>F-section</u> iff for some f or Ω with some domain U and some t_o, x_o, and t_1, $J_1 = J$. We may refer more specifically to an <u>F-section for</u> (E).

One of the main aims in the development of the theory of the behavior of (E) without uniqueness has been to determine what can be said about F_o and J. Kneser, Fukuhara, Kamke, and Pugh all studied (E). However, their results and the theorems in this section are also true for (C).

Fukuhara proved Theorem 3.2. for (E). The following slightly more general result is given in [13] in which it is not assumed that $x_o = \Psi(t_o)$ implies $t_2 \in D_\Psi$. <u>If</u> $(t_2,x_2) \in \partial F_o$, <u>then there is a solution</u> ϕ, $\phi(t_2) = x_2$, <u>such that if</u> t <u>is between</u> t_o <u>and</u> t_2, <u>and</u> $t \in D_\phi$, <u>then</u> $(t,\phi(t)) \in \partial F_o$. For a given $(t_2,x_2) \in \partial F_o$, say $t_2 > t_o$, it can occur that for every ϕ such that $\phi(t_2) = x_2$ and every $t > t_2$, if $\phi(t)$ is defined, then $(t,\phi(t)) \in$ interior F_o. From this generalization of Fukuhara's theorem, we can prove the following corollary; (see [13, Theorem 7.7]). <u>There exists a solution</u> ϕ <u>such that</u> $\phi(t_o) = x_o$ <u>and</u>

$$(t,\phi(t)) \in \partial F_o \quad \underline{\text{for all}} \quad t \in D_\phi \ .$$

Even if J is homeomorphic to a solid disc for all $t_1 \neq t$, there may be only one such solution.

Kneser and Kamke proved that J (for (E)) was compact and connected. J may be a circle S^1 so we cannot expect J to be simply connected.

Substantial further results, announced in [3] and proved in [4], have been given by Pugh. He showed that if J_1 is a compact set such that $R^d - J_1$ is diffeomorphic to $R^d - \{0\}$, then J_1 is a section for (E). But since S^1 is an F-section, not all F-sections are of this type. He shows that all finite polyhedra are F-sections but he gives an example with an G-section that is not arc-wise connected. In his example, the following set is an F-section.

$$P_1 = \{(t,0) : |t| \leq 1\} \cup \{(\sin t, t^{-1}) : t \in [1,\infty)\} .$$

Perhaps at this point we should optimistically conjecture all compact connected sets are F-sections; however, Pugh destroys our hopes. He shows that the following set P_0 is <u>not</u> an F-section. Let $S^1 = \{(x_1,x_2) \in R^2 : x_1^2 + x_2^2 = 1\}$.

$$P_0 = S^1 \cup \{([1 - e^t] \sin t, [1 - e^t] \cos t) : t \leq 0\} .$$

The compact connected set P_0 is a circle with an arc spiraling to the circle. Pugh says in [4] that his primary result is that P_0 is not an F-section. This result can be achieved via tha study of the space of solutions. With this introduction we now define, following Hu [8],

<u>Definition 4.1.</u> A metric space X is an <u>absolute neighborhood retract</u> (ANR) if whenever a closed subset X_0 of a metric space Y is homeomorphic to X, then X_0 has a neighborhood $N \subset Y$ with a continuous function $P_N : N \longrightarrow X_0$ such that $P_N(y) = y$ for all $y \in X_0$.

Examples of ANRs include a point, a circle, a figure eight, R^d, $\{x \in R^d : |x| \leq 1\}$, and any manifold in R^d. Since all ANRs are arc-wise connected, P_0 and P_1 are not ANRs.

<u>Definition 4.2.</u> A function $\mu : Y \longrightarrow Z$, where Y and Z are topological spaces, is <u>contractible</u> if there is a continuous function $m : [0,1] \times Y \longrightarrow Z$ such that $m(0,y) = \mu(y)$ and $m(1,y) \equiv$ constant for all $y \in Y$. We say $\mu : Y \longrightarrow Y$ is <u>homotopic to the identity</u> if there is a continuous function $m : [0,1] \times Y \longrightarrow Y$ such that

$m(0,y) = \mu(y)$ and $\mu(1,y) = y$ for all $y \in Y$.

Definition 4.3. We say the topological space Y is co-contractible if for each ANR X and for each continuous $\mu : Y \longrightarrow X$, the function μ is contractible.

The above concept does not seem standard in topology. Contractibility and co-contractibility for a space are closely related.

Proposition 4.1. If Y is contractible, then it is co-contractible. If Y is a co-contractible ANR, then Y is contractible.

We now state the main result of this section improving Theorem 1.2.

Theorem 4.1. If $Q \subset U$ is co-contractible, and in particular if Q contains only one point, then S_*Q is co-contractible.

We say Y is an image of X if for some continuous $\mu : X \longrightarrow Y$, $\mu(X) = Y$. The image of a connected, compact, or arc-wise connected space X is always connected, compact, or arc-wise connected, respectively. The image of a co-contractible (or contractible) set need not be contractible or co-contractible as is shown by the following simple example. IF $X = [0,1]$ then X is contractible, but the function given by $t \in [0,1] \longrightarrow (\sin 2\pi t, \cos 2\pi t) \in S^1$ shows S^1 is the image of a contractible set $[0,1]$. By Lemma 2.1 every F-section is the continuous image of $S^*(t_o,x_o)$ under the mapping $(t_o,\phi) \longrightarrow \phi(t_1)$. Hence we may say: Every F-section is the image of a compact co-contractible space S^*Q.

Theorem 4.2. P_o is not the image of a co-contractible space.

Definition 4.4. We say Y is homotopically equivalent to Z, $Y \overset{H}{=} Z$, if there are continuous functions $h_1 : Y \longrightarrow Z$ and $h_2 : Z \longrightarrow Y$ such that the functions $h_2 \circ h_1 : Y \longrightarrow Y$ and $h_1 \circ h_2 : Z \longrightarrow Z$ are homotopic to the identity.

The sets E^d and $[0,1]$ and any contractible space are homotopically equivalent. Also an open disc with a point removed $\{(x,y) : 0 < x^2 + y^2 < 1\}$ is homotopically equivalent to S_1. If C is contractible, then $C \times X \overset{H}{=} X$.

Theorem 4.3. *If* $P \overset{H}{\equiv} P_0$, *then* P *is not an F-section.*

Theorem 4.3 was given by Pugh as an unsolved problem. The following question is now relevant. What other compact connected sets are not the images of co-contractible sets?

We make the following conjecture, which departs from our topic of solution spaces. It would be interesting if true, and would simplify some of Pugh's results which are stated in terms of "stable" funnels.

Conjecture. Let $J_1 \subset R^d$ be compact and connected and let $R^d - J_1$ be diffeomorphic to $R^d - \{0\}$. Then J_1 is an F-section for (E) for some $f : R \times R^d \longrightarrow R^d$ such that solutions of (E) are unique in a neighborhood of every point except (t_0, x_0).

Note that if $D_\phi = R$ for all solutions ϕ and if solutions are unique at (t_0, x_0), then $R^d - \{0\}$ is homeomorphic to $R^d - J$.

BIBLIOGRAPHY

[1] P. Hartman, Ordinary Differential Equations, Wiley, 1964

[2] Masuo Fukuhara, Sur une généralization d'un théorème de Kneser, Proc. Japan. Acad. 29(1953), 154-155

[3] Charles C. Pugh, Cross-Sectionsof Solution Funnels, Bull. Am. Math. Soc. 70(1964), 580-583

[4] Charles C. Pugh, Cross-Section of Solution Funnels, unpublished mimeographed notes

[5] N.P. Bhatia and G.P. Szegö, Dynamical systems: stability theory and applications, Lecture Notes in Mathematics No. 35, Springer-Verlag 1967

[6] E. Roxin, On generalized dynamical systems defined by contingent equations, J. Diff. Equations, 1(1965), 188-205

[7] George R. Sell, Non-autonomous differential equations and topological dynamics, Trans. A.M.S. 127(1967), 241-283

[8] Sze-Tsen Hu, Theory of retracts, Wayne State Univers. Press, Detroit, 1965

[9] Otomar Hajek, Theory of processes, to be published

[10] Taro Ura and Ikuo Kimura, Sur le courant exterieur à une région invariante, Comment. Math. Univ. sancti pauli Tokyo 8(1960),23-39

[11] V.V. Nemytskii and V.V. Stepanov, Qualitative Theory of Differential Equations, Princeton Univ. Press, 1960

[12] Aaron Strauss and James A. Yorke, On asymptotically autonomous differential equations, Mathematical Systems Theory, 1(1967), 175-182

[13] James A. Yorke, Invariance for Ordinary Differential Equations, to appear

[14] E. Kamke, Zur Theorie der Systeme gewöhnlicher Differentialgleichungen, II, Acta Math. 58(1932), 57-85

Part IV Special Applications

Feedback and the problem of Stability in a Single Market

Martin J. Beckmann

1. Introduction. A system is said to be in an equilibrium when there is no tendency for a change. The equilibrium is stable when this equilibrium will be reached from every initial position in a certain neighborhood of the equilibrium. Now, stability is a property of system movements that is of the adjustment processes. As we admit more general types of behavior the chance of instability is increased.

Economic theory has primarily considered two types of adjustment processes which may be described in terms of linear differential or difference equations with constant coefficients. In the case of differential equations there are two basic types, viz. price or quantity adjustment. According to Walras price moves in proportion to excess demand

$$\dot{P} = k(N - M)$$

where P price, $N = \nu + nP$ demand function, $M = \mu + mP$ supply function, and k the speed of adjustment.

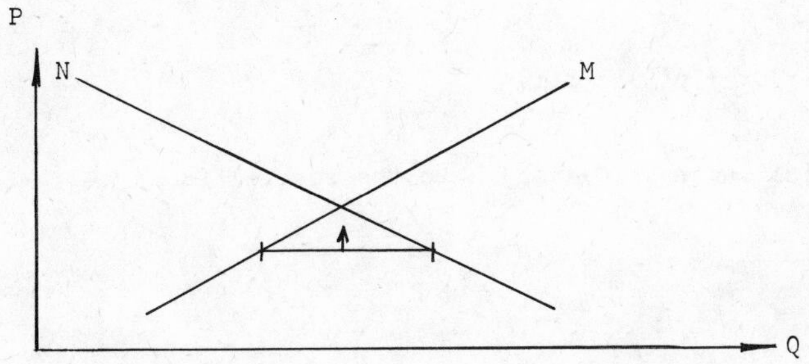

figure 1

The usual procedure is to consider supply and demand as linear functions of price in the neighborhood of the equilibrium so that a linear differential equation with constant coefficients results whose constant term may be removed by considering deviations from eqilibrium

(1) $\dot{p} = a\,p$ where $a = k \cdot (n-m)$.

The condition of stability is then, that $a < 0$.

2. Feedback. The general character of adjustment processes is revealed better when one considers it as a Feedback process (cf. A. Tustin (1953), who has applied the theory of feedback to the analysis of business cycles, also A.W. Phillips (1954)).

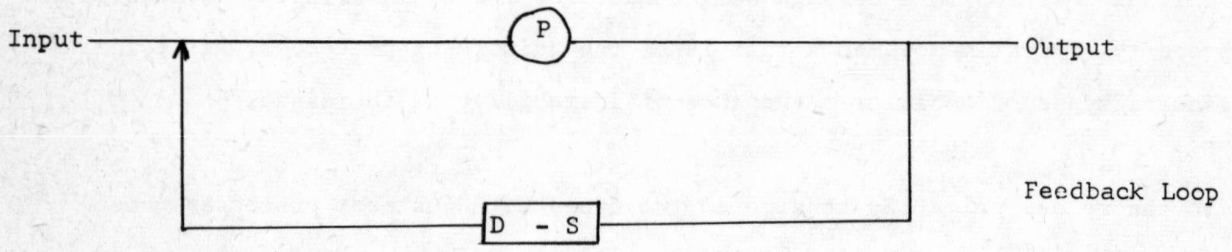

figure 2

Equation (1) in intregated from is

(1') $$p(t) = p(0) + a\int_0^t p(\tau)d\tau\ .$$

In a general feedback process passed values of price deviations occur with a general weight function

(2) $$p(t) = \int_0^t a(\tau)\,p(t-\tau)d\tau + p(0) \cdot f(t)$$

The weight f of the initial value $p(0)$ may have the special form

$$f(t) = \int_t^\infty a(\tau)\,d\tau$$

The feedback process (2) is thus characterized by a general weight function or distributed lag $a(t)$. Equation (1') shows that the Walras process is a special case of (2).

3. Criterion of Stability for price adjustment. What is the general stability criterion for adjustment processes, i.e.
what are necessary and sufficient conditions for deviations $p(t)$ to converge to zero as $t \to \infty$. Using the Laplace transform

$$X(s) = \int_0^\infty e^{-st}\, x(t)\, dt$$

equation (2) is converted to

$$P(s) = A(s) \cdot P(s) + p(o) \cdot F(s)$$

$$= p(o) \cdot \frac{G(s)}{1-A(s)}$$

A necessary condition for stability is in particular that $P(s)$ is bounded for all positive s. For arbitrary weight functions $g(t)$ this is possible only when

$$A(s) \neq 1 \qquad \text{all} \quad s > 0.$$

Since $\lim_{s\to\infty} A(s) = 0$ it follows that

$$A(s) < 1 \qquad \text{for all} \quad s > 0.$$

The properties of $A(s)$ in the neighborhood of $s = 0$ is determined by the so-called final value theorem of Laplace transforms.

$$\lim_{t\to\infty} p(t) = \lim_{s\to 0} s\, P(s)$$

as is easily verified trough integration by parts. Stability requires that

$$\lim_{s \to 0} s\, P(s) = 0$$

and therefore

$$\lim_{s \to 0} \frac{s\, F(s)}{1-A(s)} = 0 \; .$$

For example: from equation (1')

$$p = a \int_0^t p(\tau)\, d\tau + p(0)$$

one has

$$A(s) = \frac{a}{s} \qquad\qquad F(s) = \frac{1}{s}$$

Now, $A(s) < 1$ for all $s > 0$ if and only if $a \leq 0$. In view of $F(s) = \frac{1}{s}$ one has

$$\lim_{s \to 0} sP(s) = p_0 \; s \cdot \frac{1/s}{1-a/s} = p \cdot \frac{s}{s - a} = 0$$

if and only if $a \neq 0$, therefore $a < 0$. This is in agreement with the above stability criterion for Walrasian price adjustment.

4. Simultaneous Price and Quantity Adjustment. For long periods not price but quantity is the natural variable of adjustment. According to Marshall the quantity supplied moves in proportion to the excess of demand price over supply price

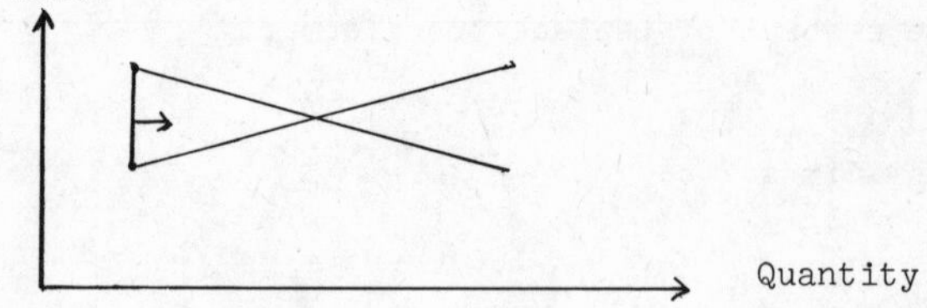

figure 3

The linear differential equation which is then obtained, using our previous notation for supply and demand functions is again of type (1)

$$\dot{q} = h\left(\frac{1}{n} - \frac{1}{m}\right) \cdot q \,.$$

The above analysis may therefore be carried out in the same way.

More interesting is the consideration of simultaneous price and quantity adjustments

$$\text{(4)} \qquad \begin{aligned} \dot{P} &= k[v + m \cdot P - Q] \\ \dot{Q} &= h\left[P + \frac{\mu}{n} - \frac{1}{n} Q\right] \end{aligned}$$

Here the first equation says that price moves in proportion to the excess of quantity demanded over quantity available.

The second equation states that the quantity supplied changes in proportion to the difference between existing price and supply price (i.e. the price at which the quantity now supplied would just be offered). The general form of price quantity adjustment processes in terms of deviations from equilibrium is

$$\text{(5)} \qquad \begin{aligned} \dot{p} &= a\,p + b\,q \\ \dot{q} &= c\,p + d\,q \,. \end{aligned}$$

Now it may easily be seen that the stability condition is that the eigenvalues of the system, i.e. the roots λ of

$$\det \begin{vmatrix} a - \lambda & b \\ c & d - \lambda \end{vmatrix} = 0$$

have negative real parts.

5. <u>Expectations and Stock Variables.</u> The previous considerations may be furhter extended by introducing price expectations in the supply or demand equations, namely whenever the time of decisions diverges from the time at which the results of the decision reached the market (a famous example is the so-called cobweb). Now, expectations are formed on the basis of experiences in the past, and this in turn introduces a feedback mechanism of type (2) above.

In addition stocks may be introduced as variables and the mechanism of price adjustment may involve changes of stock. In this case (2) it is easily shown that the feedback mechanism is involved, so that all special cases discussed so far may be represented in terms of a more general type of linear processes where simultaneously quantity and prices are moving. This assumes the form

$$p(t) = \int_0^t a(\tau)\, p(t-\tau)\, d\tau + f(t) \cdot p(0)$$

$$+ \int_0^t b(\tau)\, q(t-\tau)\, d\tau + g(t) \cdot q(0)$$

$$q(t) = \int_0^t c(\tau)\, p(t-\tau)\, d\tau + h(t) \cdot p(0)$$

$$+ \int_t^\infty d(\tau)\, q(t-\tau)\, d\tau + k(t) \cdot q(0)$$

Upon application of the Laplace transform we have

$$P(s) = A(s) \cdot P(s) + B(s) \cdot Q(s) + p(0) \cdot F(s) + q(0) \cdot G(s)$$

$$Q(s) = C(s) \cdot P(s) + D(s) \cdot Q(s) + p(0) \cdot H(s) + q(0) \cdot K(s)$$

or in matrix form

$$(6)\qquad \begin{pmatrix} P \\ \\ Q \end{pmatrix} = \begin{pmatrix} 1 - A & -B \\ \\ -C & 1 - D \end{pmatrix} \begin{pmatrix} p_0 \cdot F + p_0 G \\ \\ p_0 \cdot H + q_0 K \end{pmatrix}$$

. <u>Laplace Transformability of Solutions.</u> A first necessary condition for stability is that finite solutions P,Q exist for every admissible s , i.e. for every s with positive real part Re (s) > 0 since now, in general, complex valued solutions will occur. For this in turn it is necessary that

$$(7)\quad \det \begin{vmatrix} 1-A(s) & -B \\ \\ -C & 1-D(s) \end{vmatrix} \neq 0 \qquad\qquad \text{all} \quad s \quad \mathrm{Re}(s) > 0.$$

The solution of (6) is

$$P(s) = \frac{1}{\det} \cdot \left[(1-D)(p_0 F + q_0 G) + C(p_0 H + q_0 K)\right]$$

$$Q(s) = \frac{1}{\det} \cdot \left[B(p_0 F + q_0 G) + (1-D)(p_0 H + q_0 K)\right]$$

where

$$\det = \begin{vmatrix} 1-A(s) & -B(s) \\ \\ -C(s) & 1-D(s) \end{vmatrix} \quad .$$

For this we shall also write

$$\begin{pmatrix} P \\ Q \end{pmatrix} = \frac{1}{\det} \cdot \begin{pmatrix} \phi(s) \\ \Psi(s) \end{pmatrix}$$

7. Convergence to Zero. If $x(t)$ is Laplace transformable then again

$$\lim_{s \to 0} sX(s) = \lim_{t \to \infty} x(t) .$$

From this as a second necessary condition we obtain that

$$0 = \lim_{s \to 0} s\, P(s) \equiv \lim_{s \to 0} \frac{s \cdot \phi(s)}{\det}$$

and accordingly

$$0 = \lim_{s \to 0} s\, Q(s) = \lim_{s \to 0} \frac{s \cdot \Psi(s)}{\det} .$$

It follows that for stability either

$$\lim_{s \to \infty} \det = \infty$$

and

A, B, C, D, F, G, H, K

are bounded or

det is bounded and

$$\lim_{s \to 0} s\, \phi(s) = 0$$

$$\lim_{s \to 0} s\, \Psi(s) = 0$$

or that the numerator vanishes with a higher order than the denominator or converges to infinity with a lower order than the denominator.

These conditions may be summarized by stating that solutions of (6) must be bounded for all s with $Re(s) > 0$ and for $s \to 0$ must be of smaller order than $\frac{1}{Re(s)}$.

By the general theory of Laplace transform these conditions are also sufficient for stability.

In the following we shall apply them to some standard cases.

. Constant Coefficients. One has

$$A(s) = \frac{a}{s} \qquad B(s) = \frac{b}{s}$$

$$C(s) = \frac{c}{s} \qquad D(s) = \frac{d}{s}$$

$$F(s) = G(s) = H(s) = K(s) = \frac{1}{s}$$

The first condition is that

$$\begin{vmatrix} 1 - \frac{a}{s} & - \frac{b}{s} \\ - \frac{c}{s} & 1 - \frac{d}{s} \end{vmatrix} \neq 0 \qquad \text{for all } s \text{ with } Re(s) \geq 0 .$$

For $s \neq 0$ it follows that

$$(s-a)\cdot(s-d) - bc \neq 0 \qquad \text{all } s \text{ with } Re(s) > 0.$$

This is also the condition that no eigenvalues of positive real part exist. The second condition requires that

$$\lim_{s \to 0} \frac{s}{\det} \cdot \left[(1- \frac{d}{s})(\frac{p_0}{s} + \frac{q_0}{s}) + \frac{c}{s} (\frac{p_0}{s} + \frac{q_0}{s})\right] = 0$$

$$= \lim \frac{1}{\begin{vmatrix} s-a & -b \\ -c & s-d \end{vmatrix}} \left[(s-d)(p_0+q_0) + c(p_0 + q_0)\right]$$

$$= \frac{(p_0+q_0)s^2 + (c-d)(p_0+q_0)s}{ad - bc} = 0$$

provided that also for $s = 0$ $\det \neq 0$. That means that eigenvalues with zero real parts are not admissible either, the same applies to $q(t)$.

9. Exponential Weight Functions. Assume

$$a(t) = a \cdot e^{-rt}$$

$$f(t) = \int_t^\infty a \cdot e^{-r\tau} d\tau = \frac{a}{r} e^{-rt}$$

$$b(t) = b \cdot e^{-rt}$$

$$g(t) = \frac{b}{r} \cdot e^{-rt} \qquad \text{etc.}$$

These system of integral equations

$$p(t) = a \int_0^t e^{-r(t-\tau)} p(\tau)d\tau + b \int_0^t e^{-r(t-\tau)} q(\tau)d\tau$$

$$+ \frac{a}{r} e^{-rt} + \frac{b}{r} e^{-rt}$$

$$q(t) = c \int_0^t e^{-r(t-\tau)} p(\tau)d\tau + d \int_0^t e^{-r(t-\tau)} q(\tau)d\tau$$

$$+ \frac{c}{r} e^{-rt} + \frac{d}{r} e^{-rt}$$

is transformed upon multiplication by e^{rt}, differentiation, and introduction of new variables

$$e^{rt}p(t) = u(t)$$

$$e^{rt}q(t) = v(t)$$

into the system of differential equations

$$\dot{u}(t) = a\,u(t) + b\,v(t)$$

$$\dot{v}(t) = c\,u(t) + d\,v(t)\ .$$

The solutions p, q will approach zero if and only if u, v grow less rapidly than e^{rt}. For this it is necessary and sufficient in turn that the real parts of the eigenvalues λ of

$$\begin{vmatrix} a-\lambda & b \\ c & d-\lambda \end{vmatrix} = 0$$

are smaller than r.

In order to apply the above criterion we observe that

$$A(s) = \frac{a}{r+s} \qquad B(s) = \frac{b}{r+s}$$

$$C(s) = \frac{c}{r+s} \qquad D(s) = \frac{d}{r+s}$$

$$F(s) = \frac{a}{r}\cdot\frac{1}{r+s} \qquad G(s) = \frac{b}{r}\,\frac{1}{r+s}$$

$$H(s) = \frac{c}{r}\cdot\frac{1}{r+s} \qquad D(s) = \frac{d}{r}\,\frac{1}{r+s}\ .$$

The first necessary condition is

$$\det \begin{vmatrix} \frac{a}{r+s} - 1 & \frac{b}{r+s} \\ \frac{c}{r+s} & \frac{d}{r+s} - 1 \end{vmatrix} \neq 0 \qquad \text{all } s \ \ Re(s) > 0$$

Since for positive r, $r + s \neq 0$ this is equivalent to

$$\det \begin{vmatrix} a-(r+s) & b \\ c & d-(r+s) \end{vmatrix} \neq 0 \ .$$

This means that no eigenvalue of any real part must be greater than r. The second necessary condition

$$\lim_{s\to 0} \frac{s}{\begin{vmatrix} \frac{a}{r+s} -1, & \frac{b}{r+s} \\ \frac{c}{r+s}, & \frac{d}{r+s} -1 \end{vmatrix}} \cdot \left[-\left(\frac{d}{r+s} - 1\right) \cdot \frac{a+b}{r(r+s)} + \frac{c}{r+s} \cdot \frac{c+d}{r(r+s)}\right] = 0$$

and in the same way for $\lim_{s\to 0} s\, Q(s)$ is satisfied, provided that for $s = 0$ the determinant is unequal zero.

If the exponents of the various weight functions $a(t)$, $b(t)$, $c(t)$, $d(t)$ are different, one obtains corresponding criteria by replacing r with r_i, $i = 1, \dots 4$.

10. Integrable Weight Function. Assume

$$\int_0^\infty a(t)\,dt = a < \infty \quad \text{and} \quad f(t) = \int_t^\infty a(\tau)\,d\tau \quad \text{etc.}$$

1st condition

$$(7) \qquad \begin{vmatrix} A(s)-1 & B(s) \\ C(s) & D(s)-1 \end{vmatrix} \neq 0.$$

Since $\lim_{s \to 0} A(s) = 0$ one has

$$\det = \begin{vmatrix} -1 & 0 \\ 0 & -1 \end{vmatrix} = 1 .$$

For $s = 0$ $A(s) = a$ etc. Therefore in particular

$$\begin{vmatrix} a-1 & b \\ c & d-1 \end{vmatrix} > 0.$$

<u>2nd condition</u>

One has

$$F(s) = \int_0^\infty e^{-st} \int_t^\infty a(\tau)d\tau \, dt$$

$$= -\frac{e^{-st}}{s} \int_t^\infty a(\tau)d\tau \Big|_0^\infty + \int_0^\infty \frac{e^{-st}}{s} \cdot (-a(t))dt$$

$$= \frac{a}{s} - \frac{1}{s} A(s)$$

$$sP(s) = \frac{s}{\begin{vmatrix} A(s)-1 & B(s) \\ C(s) & D(s)-1 \end{vmatrix}} \cdot \Big[-p_0(D(s)-1) \cdot \frac{1}{s}(a+b-A(s)-B(s)) + q_0 \cdot C(s) \cdot \frac{1}{s}(c+d-C(s)-D(s))\Big]$$

$$\lim_{s\to 0} sP(s) = \frac{1}{\begin{vmatrix} a-1 & b \\ c & d-1 \end{vmatrix}} \cdot |-p_0(d-1)(a+b-a-b) + q_0 \, c(c+d-c-d)|$$

provided that also for $s = 0$ the determinant does not vanish. The conditions of stability are therefore (7) for all s with $Re(s) \geq 0$.

11. <u>Conclusions.</u> The problem of stability in economics can never be settled definitively but only with reference to a class of processe. But more general processes may always be introduced. For instance it may be desirable to consider non-linear adjustment processes in markets. Assume for instance that all firms of an industry have a supply function resulting from linear programs involving the same cost levels.

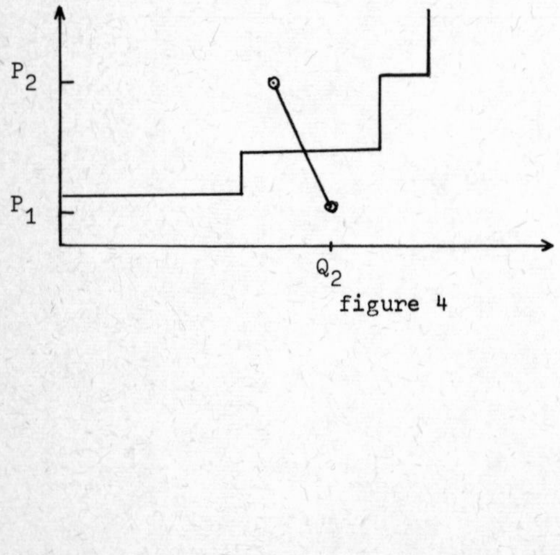

figure 4

The supply curve of an industry is then also a step function with these cost levels. Assume that this demand curve intersects as drawn. Then a cobweb will result such that prices will move between P_1 and P_2 and the quantity supplied between zero and Q_2 . In this case it is not admissible to replace the supply curves in the neighborhood of an equilibrium by a linear supply curve: for the amplitudes are necessarily so large that one exceeds the range of linearity. Here - as in the stability analysis of systems other than markets (production programs, inventory systems, queues) is a wide area still open for further research.

References

|1| Tustin, A., The Mechanism of Economic Systems, London 1953

|2| Phillips, A.W., Stabilisation Policy in a Closed Economy, Economic Journal, 64 (1954) 290-323.

INVESTIGATIONS ON ORGANIZATION OF PRODUCTION PROCESSES WITH TREE STRUCTURE *

Andrzej J. Blikle

As is suggested by the title we shall be concerned here with production processes having tree-structure. First we shall introduce some basic notions of the theory of graphs. As there are no proofs in this paper (they can be found in references) our definitions need not be formulated in a very formal way.

By a graph we shall mean a pair $\langle G,E \rangle$ where G is a finite set of objects called points or vertices of the graph and E is a collection of ordered pairs (not necessarily all) of the elements of G, thus $E \subseteq G \times G$. Usually graphs will be represented pictorally as a set of points on the plane (the set G) where some points are joined one with the other by continuous lines with arrowheads. Two points a and b in G joined together with an arrowhead at b represent an ordered pair $\langle a,b \rangle$ in E. The point a is then called a predecessor of b and b is called a successor of a. An example of a graph is given in fig. 1 where $G = \{a_1,\ldots,a_8\}$.

$E = \{\langle a_1,a_3 \rangle, \langle a_1,a_2 \rangle, \langle a_3,a_4 \rangle, \langle a_3,a_5 \rangle, \langle a_5,a_4 \rangle, \langle a_7,a_8 \rangle\}$ but for instance $\langle a_3,a_1 \rangle$ does not belong to E. The point a_1 is a predecessor of a_3 and a successor of a_5.

---

* Results stated in this paper have been developed for the use in the theory of addressless computers. Principal ideas of these results were introduced by Z. PAWLAK in many papers - among them [4], [5], [6] - and a mathematical discussion has been given by the author of the paper in [3].

The possibility of application of those results in economics is based on the fact that processes of production and processes of computation are frequently very similar.

Figure 1

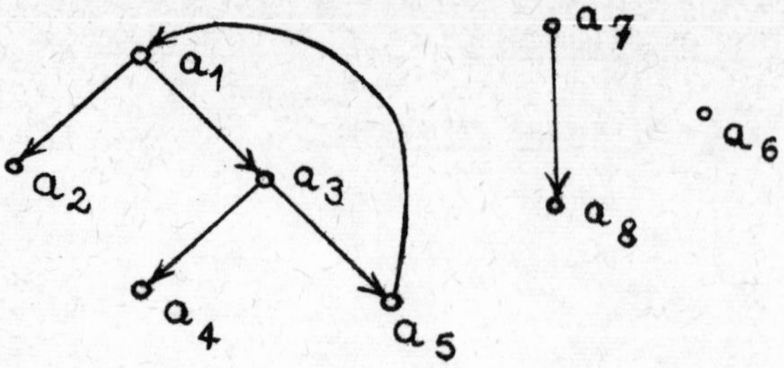

By a chain in a given graph <G,E> we shall mean any sequence $g_1,\ldots,g_n$ of points in G such that for every $1 \leq i \leq n-1$ either $\langle g_i,g_{i+1}\rangle$ or $\langle g_{i+1},g_i\rangle$ is in E. E.g. the sequence a_2,a_1,a_3,a_5 is a chain for the graph in fig. 1.

A graph <G,E> is said to be connected if for any two points g_i,g_j in G there exists a chain which starts at one of them and ends at the other. E.g. the graph in fig. 1 is not connected because there does not exist a chain between a_1 and a_7. The graph composed of points $a_1,\ldots,a_5$ is connected.

A chain $g_1,\ldots,g_n$ of a graph is said to be a cycle if all its points are different except $g_1 = g_n$. E.g. a_1,a_3,a_5,a_1 on fig. 1 is a cycle.

By a tree we shall mean any connected graph <G,E> without a cycle which satisfies the following two conditions:

(i) every point in G has at most one successor;

(ii) there exists exactly one point in G without a successor. This point is called the root of the tree.

Points of a tree which have predecessors are called nodes. The others are called free points.

An example of a tree is given in fig. 2. The points $b_1,\ldots,b_6$ are nodes, point b_1 is a root and $b_7,\ldots,b_{12}$ are free points.

Figure 2

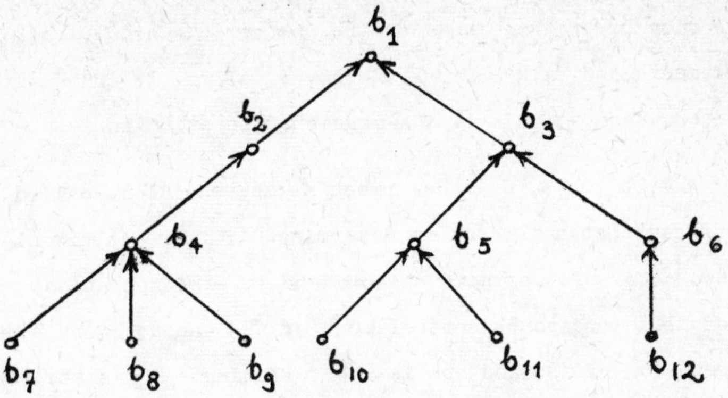

Let us consider now, as an example, a simple process of production with a tree structure. Let it be a process of production of a sandwich (S) out of two slices of bread (Br), two pieces of butter (Bu) and two slices of ham (H). The corresponding tree is following:

Figure 3

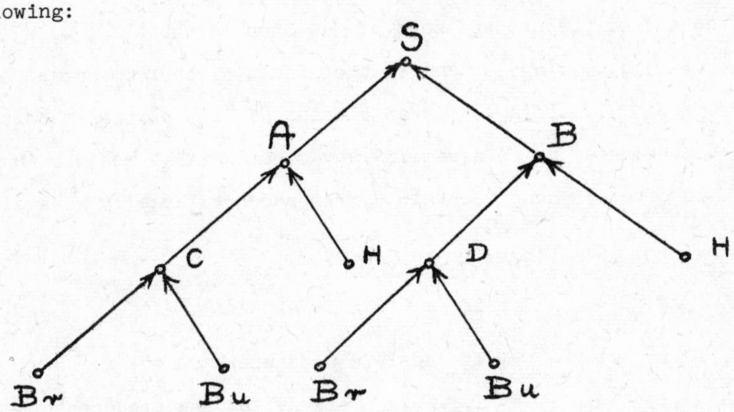

As one can see, free points of the tree correspond to prime components (initial products) of the sandwich and nodes correspond to the operation of "putting together". In fact, nodes always represent operations of a considered process and free points always represent some initial products. We shall introduce now a formal notation which will be used later to describe this fact.

Suppose that we are given a tree $\mathfrak{A} = \langle G,E \rangle$ and let N and F be respectively the set of nodes and the set of free points of $\mathfrak{A}$. Thus $N \subseteq G$, $F \subseteq G$ and $G = N \cup F$. Let there be given now two functions ϕ and ψ where:

ϕ - is a function defined in F and with values in some set P

ψ - is a function defined in N and with values in a set O of operations which have arguments and values in P. Moreover, if a is in N and has exactly k predecessors then $\psi(a)$ is a k-argument operation in P.

Any system $< \mathfrak{A}, P, O, \phi, \psi >$ described as above will be called a process *. To simplify, we can say that a process is determined if there is given a tree $\mathfrak{A} = <G,E>$ and two sets: P - of some objects and O - of operations in P, and if to every free point of $\mathfrak{A}$ corresponds some element of P and to every node - some element of O. The interpretation of P and O is the following: P is the set of all initial products (input products), all half-products (called partial results abv.p.r.) produced in the course of executing the process and the final product. O is the set of all operations performed during the execution of the process. It is easy to see now that some elements of P (namely partial results) correspond to the nodes. In fact, let a be a node and $a_1,\dots,a_k$ be all of the predecessors of a. Suppose that $a_1,\dots,a_k$ are in F. If $\psi(a) = \Delta$ and $\Delta(\phi(a_1),\dots,\phi(a_k))$ denotes the result of performing Δ on $\phi(a_1),\dots,\phi(a_k)$, then $\Delta(\phi(a_1),\dots,\phi(a_k))$ is an element of P (partial result) and corresponds in some way to the node a. In a similar way other partial results in P correspond to other nodes. In order to describe this correspondence formally we shall introduce a function f:

$f(a) = \phi(a)$ if a is a free point

$f(a) = \Delta(f(a_1),\dots,f(a_k))$ if a is a node where $\Delta = \psi(a)$ and $a_1,\dots,a_k$ are all of the predecessors of a.

If a is a root then f(a) is called a terminal result of the considered process. Thus, we execute a process in order to produce f(a) where a is the root of the process.

Let us consider now, as an example, a process of computing an arithmetical formula. This is perhaps not a proper example of a production-process, but on the one hand in a formal discussion we are interested rather in the structure than in the nature of a process and on the other hand, computation can be considered as a production of numbers. Let there be given thus a formula:

$$(((7+4) - (6+1)) + ((5-3) \cdot (2+1) : 2)) \qquad /1/$$

* This notion corresponds to the notion of computation in [3]

To this formula corresponds the following tree:

Figure 4

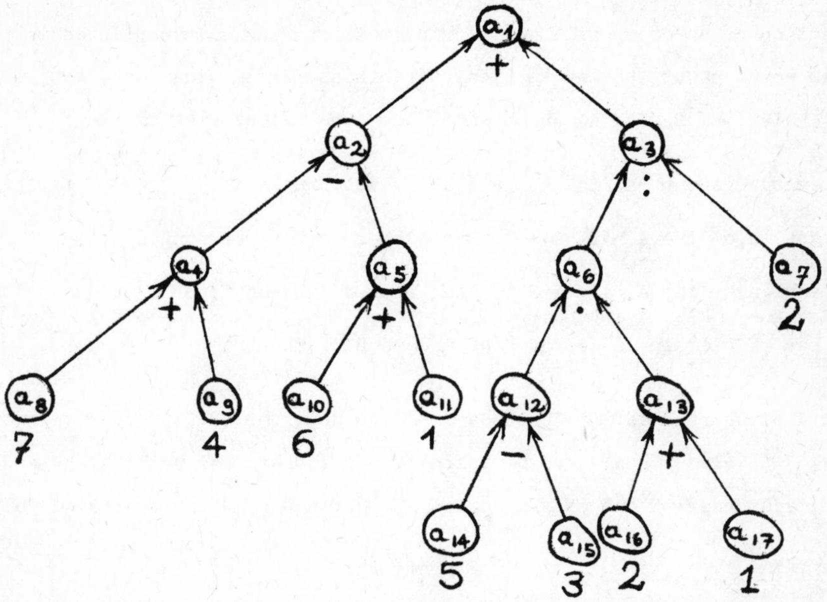

Functions ϕ and ψ are given explicitly in the figure. The sets O and P are respectively:

$$O = \{ + , - , \cdot , : \}$$

$$P = \{1, 2, 3, 4, 5, 6, 7, 11\}$$

Suppose now that at any moment we are able to perform only one operation*. This assumption requires establishment of some successiveness in which operations will be performed during execution of the process. As "successiveness" is not a mathematical notion we shall talk about an order between operations of the process, thus about an order in the set of

* It seems that this assumption can be supposed to be satisfied in many cases of production processes. Moreover, results obtained for this simple case are very helpful for discussing the case where many operations can be performed at the same moment.

nodes. Evidently such an order should satisfy the condition that if a is a predecessor of b then a is before b in this order. Orders which satisfy this condition are called admissible orders *.

As the notion of an order will be of principal importance in our investigations, we shall say a few words about its mathematical definition. By an order in a set A we mean a two-argument relation R in A which satisfies the following conditions:

1. for every element a in A, a R a is not true
2. for any a,b in A if $a \neq b$ then either a R b or b R a
3. for any a,b in A if a R b then b R a is not true
4. for any a,b,c in A if a R b and b R c then a R c

The formula a R b we read as: a precedes b in the order R. As an example of an order relation " < " (less than) can be taken in the set of real numbers. By an order converse to a given order R we mean an order $\bar{R}$ defined by: a $\bar{R}$ b if and only if b R a.

Suppose now that we are given a process $\pi = \langle \mathfrak{A}, P, O, \phi, \psi \rangle$ and some admissible order R in the set of points of $\mathfrak{A}$ **. Thus we have complete information about how to execute this process. An ordered process (a pair π, R) can be thus understood as a kind of program of execution of some physical process.

In order to execute process π we have to adopt some convention establishing a method of storing partial results. We shall assume here that partial results will be stored successively one after the other, every result labelled by a symbol of the name of the corresponding node (thus $f(a_i)$ labelled by a_i). A labelled partial result will be denoted by a pair $(a_i, f(a_i))$. This labelling will help afterwards in finding required partial results in the store.

* in [3] admissible orders are those whose converse is admissible in our sense. This follows from a difference in notation and is immaterial for the topic.

** By an order in a process we shall always mean an order in the set of all points of the corresponding tree, just to have an order not only between operations of the process but also between arguments of every operation.

Now let us turn back to our example on fig. 4. Suppose the nodes are ordered as follows:

$$a_{12}, a_5, a_{13}, a_4, a_2, a_6, a_3, a_1$$

The step-by-step execution of formula (1), corresponding to order R as given by the above sequence of nodes, can be described by the following table:

Table 1

cons. node	perf. oper.	state of store after performance of the operation
a_{12}	5 - 3	(a_{12},2)
a_5	6 + 1	(a_{12},2) (a_5,7)
a_{13}	2 + 1	(a_{12},2) (a_5,7) (a_{13},3)
a_4	7 + 4	(a_{12},2) (a_5,7) (a_{13},3) (a_4,11)
a_2	11 - 7	(a_{12},2) ~~(a_5,7)~~ (a_{13},3) ~~(a_4,11)~~ (a_2,4)
a_6	2 · 3	~~(a_{12},2)~~ ~~(a_5,7)~~ ~~(a_{13},3)~~ ~~(a_4,11)~~ (a_2,4) (a_6,6)
a_3	6 : 2	~~(a_{12},2)~~ ~~(a_5,7)~~ ~~(a_{13},3)~~ ~~(a_4,11)~~ (a_2,4) ~~(a_6,6)~~ (a_3,3)
a_1	4 + 3	~~(a_{12},2)~~ ~~(a_5,7)~~ ~~(a_{13},3)~~ ~~(a_4,11)~~ ~~(a_2,4)~~ ~~(a_6,6)~~ ~~(a_3,3)~~ (a_1,7)

Partial results which have been used obviously do not appear any more in the store. Here, in tab. 1, we do not erase them just in order to make the picture of the execution of the process more clear.

This example shows a typical way in which production-processes with tree structure are performed. The fact that partial results are stored successively (in a line) is evidently immaterial if we have labels. The use of labels expresses in a mathematical language that we have a device or ability for recognizing at every moment appropriate partial results in the store. In fact, labels are frequently used for this aim in practice.

This way of execution of processes, however typical, has several inconveniences:

1. It needs, as mentioned above, ability for recognizing appropriate partial results.
2. During the whole time of execution of the process we have to observe the tree in order to be able to establish which of the partial results is required at a given moment.
3. In every step we have to search through the store in order to find the required result. This is especially inconvenient when the number of p.r. in store is large.

As it turns out, all of those difficulties can be avoided by using some special orders to execute processes. We shall start with an example of one of them. This order is denoted by $\bar{p}$ and has been proposed by Z. PAWLAK. It can be described informally in the following way:

First we enumerate all points of a given tree in such a manner that the root has number one and the other points are numbered level by level in every level from the left to the right. An example of such enumeration is given in fig. 4 where corresponding numbers of points are simply their indices. The order $\bar{p}$ corresponds now to an order converse to the one given by the above described enumeration. Thus $a_i \; p \; a_k$ if and only if $i > k$. The string:

$$a_{13}, a_{12}, a_6, a_5, a_4, a_3, a_2, a_1 \qquad /2/$$

is formed of all nodes of the tree on fig. 4 ordered according to $\bar{p}$. When executing processes ordered in $\bar{p}$ we need neither labeling, nor any device for recognizing required partial results, nor searching through the store. This is a consequence of the existence of the following principle stated in [6] and proved in [3]:

If we perform any process ordered in $\bar{p}$ and if we store partial results successively one after the other from the left to the right, then at every moment:

- If we need one partial result, then it is the leftmost partial result in the store.
- If we need two partial results, then they are two leftmost partial results in the store where the first is the right and the second is the left argument of an actually performed operation*.

---

* Those rules concern the case of biargument operations, they can be, however, easily extended for the case of n-ary operations.

Here we assume evidently that every partial result after being used does not appear any more in the store.

The process given in fig. 4, when ordered in $\bar{p}$, will be executed in the following way:

Table 2

cons. node	perf. oper.	state of store after performance of the operation
a_{13}	2 + 1	(3)
a_{12}	5 - 3	(3) (2)
a_6	2 · 3	~~(3)~~ ~~(2)~~ (6)
a_5	6 + 1	~~(3)~~ ~~(2)~~ (6) (7)
a_4	7 + 4	~~(3)~~ ~~(2)~~ (6) (7) (11)
a_3	6 : 2	~~(3)~~ ~~(2)~~ ~~(6)~~ (7) (11) (3)
a_2	11 - 7	~~(3)~~ ~~(2)~~ ~~(6)~~ ~~(7)~~ ~~(11)~~ (3) (4)
a_1	4 + 3	~~(3)~~ ~~(2)~~ ~~(6)~~ ~~(7)~~ ~~(11)~~ ~~(3)~~ ~~(4)~~ (7)

Processes of production or computation are very frequently executed by automata. An automaton which performs a process should evidently know its structure, i.e. the corresponding tree. We need thus a formal language, comprehensible for automata, in which trees could be described in an unambiguous manner. In fact, many such languages are known. We shall study here one of them introduced by Z. PAWLAK and called Pawlak-notation with blank spaces. This formalism is discussed in [3] and [6].

Let there be given a process $\pi = \langle \mathfrak{A}, P, O, \phi, \psi \rangle$ and an order R in the set G of points of $\mathfrak{A}$. Let $a_1, \dots, a_n$ be a string of all nodes of $\mathfrak{A}$ ordered in a way converse to that of R , i.e. $a_i \mathrel{R} a_j$ iff $i > j$. Now every symbol a_i in the string we replace by the string of symbols

$$\ulcorner \psi(a_i) \urcorner \; \alpha_{i1} \dots \alpha_{ik}$$

where $a_{i1}, \dots, a_{ik}$ are all predecessors of a_i ordered in R and

$$\alpha_{ij} = \begin{cases} * & \text{if } a_{ij} \text{ is a node} \\ \ulcorner\phi(a_{ij})\urcorner & \text{if } a_{ij} \text{ is not a node*} \end{cases}$$

A formula which we obtain in this way is called an R-program of the process π. E.g. a $\bar{p}$-program of process given on fig. 4 is following:

$$+ ** - ** : * 2 + 7 4 + 6 1 \cdot ** - 5 3 + 2 1$$

This program we execute from the right side to the left, step by step, considering in every step a subformula of the form

$$\ulcorner\psi(a_i)\urcorner \ \alpha_{i1}\ \alpha_{i2} \cdots \alpha_{ik}$$

which we treat as the following order (command) of the program:

Perform operation $\psi(a_i)$ where for every $1 \leqslant j \leqslant k$ the j-th argument is either an initial datum $\phi(a_{ij})$ if $\alpha_{ij} = \ulcorner\phi(a_{ij})\urcorner$ or a partial result if $\alpha_{ij} = *$.

Evidently, for different orders we obtain different programs for a given process. It is proved in [3] that for every admissible order R and every process π the R-program of π describes π unambiguously. However, if we assume that programs are performed step by step from the right to the left, then for many orders principles of locating partial results in the store do not exist. It means that for many orders after reading a part of the program we are not able to locate the required partial result in the store without seeing the remaining part of program. Orders for which an appropriate principle exists are called addressless orders. As is proved in [3], for the family of trees with binary ramifications there exist exactly six addressless orders. Four of them, $\bar{p}$ in this group, have been given by Z. PAWLAK and two others by the author of this paper.

The stated result can be easily generalized for the case of trees with n-ary ramifications, and by a quite analogous way as in [3] it can be proved that for family of all trees with at most n-ary ramifications there exist exactly $(n + 1)!$ different addressless orders. All of them can be given explicitly and all of them have extremely simple algorithms of locating arguments in the store. Namely, for any addressless order if partial

---

* $\ulcorner\phi(a_{ij})\urcorner$ means: the symbol of $\phi(a_{ij})$

results are stored from the left side to the right - one after the other - then in every moment appropriate p. r. are some rightmost or some leftmost or some rightmost and some leftmost p. r., but never p. r. from the "inside" of the store.

Another example of an addressless order is an order called w. The principle of corresponding enumeration of points of a tree is following. The root is of number 1. Then we enumerate successive points passing along the tree left-down as long as possible. When this is no more possible, we go back to the nearest node, therefore we go right-down in the first step and then left-down again. An example of such enumeration is given in fig. 5.

Figure 5

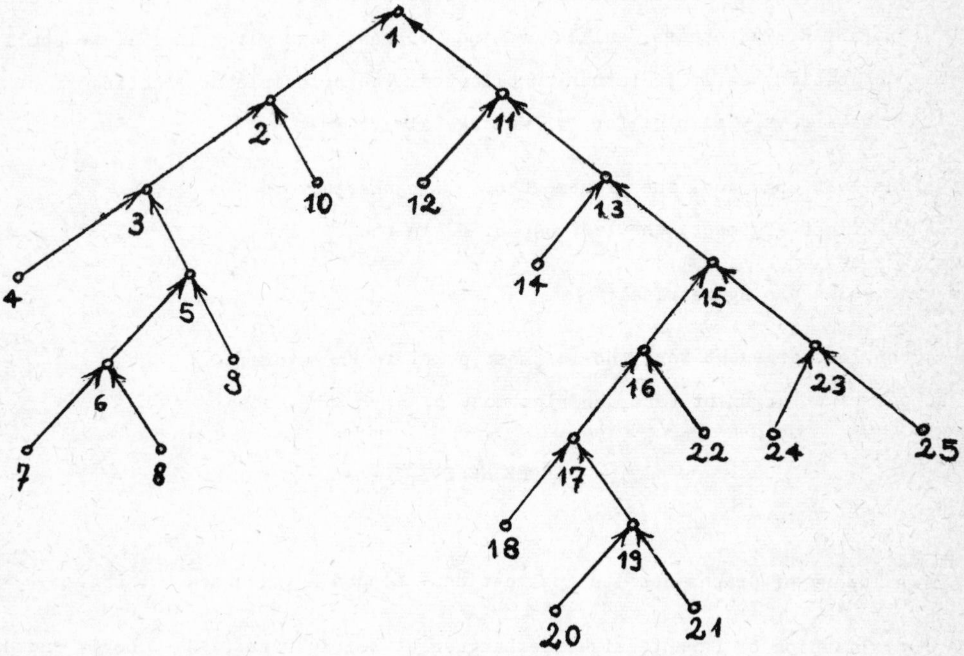

A w-program of the process given on fig. 4 is following:

+ ** - ** + 7 4 + 6 1 : * 2 · ** - 5 3 + 2 1

Now, when we execute a w-program of a process, we have a principle of locating partial results in the store to the effect that if p. r. are stored successively from the left to the right, then:

- If one partial result is required, then it is the rightmost p. r. in the store.

- If two p. r. are required then they are the two rightmost p. r. in the store, where the first is the left and the second is the right argument of the actually performed operation.

It is easy to see that replacing words 'left' by 'right' and conversely in definitions of orders $\bar{p}$ and w, we obtain definitions of orders symmetrical in some way to $\bar{p}$ and w. The new orders are called respectively p and $\bar{w}$ and are also addressless orders. Evidently, corresponding rules of locating partial results differ from those of $\bar{p}$ and w only in locating left and right arguments.

We have described here four addressless orders: p, $\bar{p}$, w, $\bar{w}$. All of them have been found by Z. PAWLAK. Two others, called v and $\bar{v}$ have been given in [3]. We shall not quote its definition, as it is rather complicated. The principle of locating p. r. in the store is, however, very simple: for v we take always

- as the left argument, the rightmost p. r. in the store;
- as the right argument, the leftmost p. r. in the store.

For $\bar{v}$ the rules are symmetrical:

- as the left argument take the leftmost p. r. in the store;
- as the right argument take the rightmost p. r. in the store.

REFERENCES

[1] BERGE, C.,
The Theory of Graphs and its Applications. London Methuen and Co. Ltd.

[2] BLIKLE, A. J.,
Formalization of Parenthesis-free Languages. Zeitschrift f. Mat. Logik und Grundlagen der Mathematik, Bd. 12 (1966), pp. 177-186.

[3] BLIKLE, A. J.,
Mathematical Investigations in the Theory of Organization of Addressless Computers (to be published).

[4] PAWLAK, Z.,
Organization of the Address-free Digital Computer for Calculating Simple Arithmetical Expressions. Bulletin de l'Académie Polonaise des Sciences, Série de Sciences Techniques, Vol. 8, 4(1960), p. 193.

[5] PAWLAK, Z.,
Realization of Memory of Partial Results in Certain Paranthesis-free Formalism. Ibid. Vol. IX, 8(1961), p. 487.

[6] PAWLAK, Z.,
Realization of Memory of Partial Results in Certain Paranthesis-free Formalism. Ibid. Vol. IX, 8(1961), p. 487.

Testing Econometric Models by Means of Time Series Analysis

Claude Hillinger

I - Introduction: This paper deals with testing a proferred econometric model in order to determine if it provides an adequate explanation of a dynamic process in an actual economy. The relevance of this topic to the general theme of the International Summer School on Mathematical Systems Theory and Economics derives from the fact that economic science will gain little from the analysis and optimization of dynamic models unless it can be established that these models provide at least a first order approximation to real world data.

The branch of economics which deals with the fitting of mathematically formulated economic models to data (parameter estimation) and with the statistical testing of such models is econometrics.[1] During the past two decades econometrics has flourished and has added substantially to our knowledge of quantitative economic relationships. While some of the key effects, such as the effect of income on consumption were quickly established, it has proved difficult to narrow the representation of these effects to precise functional forms, and to discriminate between specific hypotheses competing within the same general framework. For example, while the dependence of consumption on income, denoted the consumption function, is well established, it has not been possible to show that any particular version of it, be it that of Duesenberry, Modigliani, or Friedman, is superior to the others.

The thesis of this paper is that the ability of econometric methods to discriminate between competing hypotheses can be enhanced by a careful analysis of the dynamic properties of a given econometric model, and by a comparison of these properties with aspects of the dynamic process in the real economy which can be directly inferred from the data.

1. Three standard recent text books on econometrics are Johnston (3) which is a convenient introduction to all the principal topics of econometrics, Goldberg (1) which provides a particularly good treatment of simultaneous equation systems, and Malinvaud (4) which is notable for a highly readable introduction to time series analysis.

The analysis of a dynamic process based directly on data, without intermediation of a theoretical economic model, is most naturally carried out by means of time series analysis. A previous knowledge of time series analysis is however not required for the reading of this paper.

While major emphasis will be centered on how to test the dynamic characteristics of econometric models in general, the exposition will for the sake of definiteness be carried out in terms of a concrete model which was fitted to data by the author.[2]

II - The Derivation of the Model. The existence of a business cycle of 3 to 4 years duration, in which most of the fluctuations of output are due to fluctuations in inventory investment, has been well documented. This cycle, which has been characteristic of the post World War II economic history of the U.S. until about 1961 is usually referred to as the "inventory cycle." As early as 1941, L. Metzler published a now classic paper in which he developed a simple mathematical model in which the dependence of the output of firms on their inventory levels may lead to cyclical fluctuations.

The econometric problem was the following: Starting from the theoretical ideas of Metzler, is it possible to devise a relatively simple model which when fitted to empirical data would be capable of explaining the observed fluctuations in the economy in a quantitative sense?

A convenient starting point is the consideration of a basic equation linking the stock of inventories to output and final sales

$$I_t \equiv I_{t-1} + y_t - s_t \qquad (1)$$

This equation, which is valid for individual firms as well as for the economy as a whole, simply asserts that the end of period inventory is equal to the inventory carried over from the previous period plus this period's production less this period's sales. By writing behavioral equations for the determination of Y and S this identity becomes the basis of a closed model of the economy capable of generating fluctuations.

2. A more complete discussion of this model, particularly from an economic point of view, can be found in (2).

Derivation of the sales equation is as follows: Sales depend on expected income

$$S_t = \alpha Y_{Et} \qquad 0 < \alpha < 1 \tag{2}$$

where expected income is related to observed income according to

$$Y_{Et} = Y_{Et-1} + \varepsilon(Y_{Et-1} - Y_{Et-2}) + (1 - \varepsilon)(Y_t - Y_{Et}) \qquad 0 < \varepsilon < 1 \tag{3}$$

From these can be deduced a relation of the form

$$S_t = \beta_1 Y_t + \alpha_1 S_{t-1} + \alpha_2 S_{t-2}$$

$$0 < \beta_1 < 0.5, \; 0.5 < \alpha_1 < 2, \; -1 < \alpha_2 < 0 \tag{4}$$

The output relation is obtained by assuming that desired inventories are proportional to sales and that firms minimize a loss function in the deviation of actual from desired inventories and in the rate of change of output.

$$L_t = \delta(I_t - \theta S_t)^2 + \eta(Y_t - Y_{t-1})^2$$

$$\theta > 0 \;, \; \theta > 0 \;, \; \eta > 0 \tag{5}$$

By substituting the inventory identity for I_t in (5) and setting

$$\frac{dL_t}{dY_t} = 0$$

we derive an equation which may be expressed as

$$Y_t = \beta_2 S_t + \alpha_3 Y_{t-1} + \alpha_4 I_{t-1}$$

$$\beta_2 > 0, \; 0 < \alpha_3 < 1, \; -1 < \alpha_4 < 0, \; \alpha_3 - \alpha_4 = 1 \tag{6}$$

Following is the complete system, now presented with empirically estimated coefficients

$$S_t = \underset{(0.21)}{0.16}\, Y_t + \underset{(0.44)}{1.17} S_{t-1} - \underset{(0.25)}{0.33} S_{t-2}$$

$$R^2 = .99$$

$$Y_t = \underset{(0.17)}{0.80\ S_t} + \underset{(0.17)}{0.56\ Y_{t-1}} - \underset{(0.07)}{0.38\ I_{t-1}} \tag{7}$$

$$R^2 = .99$$

$$I_t \equiv I_{t-1} + Y_t - S_t$$

It can be verified that all of the theoretical restrictions on the coefficients are satisfied.

The steps discussed thus far in this section are typical of most econometric investigations. They involve: 1) The specification of a mathematical model suggested by a combination of prior economic theory and informal observations, a specification which includes some restrictions on the admissible values of the parameters of the model. 2) Estimation of point values of the parameters by fitting the model to data. This usually involves some variant of least squares regression. 3) Evaluation of the model by checking if the estimated parameter values fall within the prescribed intervals and by determining the closeness of the static fit of the equations to the data as expressed by the R^2 statistic.

III - Dynamic aspects of the model. Equations (7) represent a fourth order difference equation system with constant coefficients. By means of the shift operator E this system can be transformed into a system of three identical fourth order difference equations, each involving only one the variables S, Y, or I. In each case the equation is of the form[3]

$$\lambda^4 - 2.18\lambda^3 + 2.09\lambda^2 - 1.03\lambda + 0.281 = 0 \tag{8}$$

The roots of this equation are complex, and are given by

$$\begin{aligned} \lambda_1, \lambda_2 &= 0.90\ (\cos 049 \pm i \sin 049) \\ \lambda_3, \lambda_4 &= 0.59\ (\cos 1.04 \pm i \sin 1.04) \end{aligned} \tag{9}$$

3. The parameter values appearing in equation (8) were computed from a different set of estimates than those appearing in (7). This has to do with the presence of a secular trend in the data as explained in (2).

Corresponding to these roots there is a solution for each of the variables, for example for Y_t, of the form

$$Y_t = \overline{Y}_t + A_1(0.90)^t \cos(0.49t-\varepsilon_1) + A_2(0.59)^t (\cos 1.04t-\varepsilon_2) \qquad (10)$$

Here $\overline{Y}_t$ represents a secular growth trend of the economy, assumed to be determined outside the model, and the remaining two terms represent the deviation from $\overline{Y}_t$ after an initial displacement which determines the constants A_1, A_2, ε_1, and ε_2. The periods corresponding to this transient part of the solution are respectively 3.23 and 1.52 years. The transient solution is thus of the form of a two term Fourier series expansion, with a major period of approximately 3.2 years. Also important is the fact that both terms of the solution are damped.

We now proceed with a discussion of what is meant by testing the dynamic characteristics of the model. Testing any scientific theory involves an attempt at finding grounds for rejecting it. Consequently testing the dynamic aspects of the model involves a search for evidence that certain aspects of the dynamic process in the real economy differ from the corresponding aspects of the dynamic process implied by the model. This raises the following question: Since we can never be sure of the true process, but can only obtain estimates of the process based on data, what justification is there for challenging the estimates represented by the model by some other set of estimates? The answer is that important characteristics of a dynamic process, in particular its moments and from these serial correlation functions or spectral density functions, can be computed using only a minimal statistical model. Because these estimates can be made without any _economic_ hypotheses about the mechanism generating the observations, they can be used as a check on the _a priori_ assumptions which make up the maintained hypothesis of the econometric model.

An illustration of this idea is provided by the following very simple examples. Suppose that our "economic model" for explaining the variation of some variable x is

$$x_t = ax_{t-1}, \quad 0 < a < 1$$

On fitting the equation to data we obtain an estimate of a which falls within the required interval, as well as a close fit as indicated by $R^2 = 0.99$. Judged solely by these static tests the model would therefore appear to be highly successful. However a direct scrutiny of the data may reveal the presence of one or more systematic periodic components. Since a first order difference equation cannot provide a periodic solution, we have found grounds for rejecting the model out of hand, even without estimating the parameter a. Another example may be given in terms of the inventory cycle model discussed in this paper. The structure of this model is such that the time path of each variable is forced to have the same periodic components and damping factors, and may differ only in amplitude and phase. This feature is part of the a priori specification of the model and cannot be refuted by fitting the model to data. However, a direct analysis of the data may reveal, for example, a single periodic component in output but two periodic components in the sales variable. This information would be sufficient to invalidate the model.

It should be clear on the basis of what has been said thus far that even in the case where we are considering a single body of data which is used to estimate the parameters of our econometric model it may still be possible by a direct examination of the data to derive from it information contradictory to the maintained hypothesis of the model. Another case which often arises in practice is that while all of the data required for estimating the parameters of the econometric model exist for only a limited time interval, partial data sufficient for estimating some dynamic characteristics may exist outside of that interval.

The direct investigation of the data in order to infer characteristics of the underlying dynamic process may involve informal as well as formal methods, and both were used to test the dynamics of the inventory cycle model. The principal formal method is time series analysis, and in particular the computation of empirical spectral density functions. For our purpose it suffices to state that time series analysis is a method for detecting possible periodic components in an otherwise random series of observations. More specifically, the spectral density function $S(\omega)$ may be interpreted as showing the fraction of the total variance of a process which may be attributed to a periodic component with angular frequency ω.[4]

With these preliminary considerations out of the way we may turn again to the concrete example of the inventory cycle model. The model had been fitted to quarterly

U.S. data for the period 1953-1961. A relevant aspect of the economic history of this period is that there was extensive speculation in inventories around 1951 due to anticipated shortages and rising prices associated with the Korean War. As a result the ratio of inventories to sales was driven to record levels. The subsequent reaction led to a recession which was followed by two more business cycles, the last one terminating in 1961. Since 1961 the variables have deviated little from their secular growth trends. The I/S ratio itself oscillated about its mean value during these cycles but these oscillations are contained within an exponentially narrowing band.[5] Economic history is thus in accord with the hypothesis that the business cycles which occurred were the manifestations of a damped transient resulting from the initial displacement caused by the Korean War and dying out about 1961.

It will be recalled that the model as estimated implies that the transient solution can be represented as a two Fourier series with a major period of 3.2 years and a half period, and that furthermore this representation must apply to all of the three variables included in the model. In order to test these aspects of the model, the three time series used were subjected to spectral analysis. Each of the resulting spectra

4. Time series analysis has recently enjoyed a minor vogue among economists. However, the initial expectations based on the use of these mathematical tools were for the most part disappointed and were followed by a considerable amount of disillusionment. Those who feel that time series analysis is of small potential importance for economics usually point out that it involves "measurement without theory." In so far as they criticize the mere computation of empirical spectra for economic time series, without making contact with a formally specified economic model, they have a valid point. What seems to have been overlooked however is that the very lack of any economic assumptions in time series analysis makes it possible to use these estimates to check the validity of a proposed economic model. This application of time series analysis lies at the heart of the present paper. The testing of econometric models represents, in the writer's opinion, the most significant area of application for time series analysis in econometrics.

5. See Fig. 1 of (2).

exhibited a peak of duration in the 3-4 year range, and a secondary peak at approximately half that duration. Because of the shortness of the empirical record which covered barely 3 cycles, the peaks were not sharply localized, however, given this data limitation, the spectral estimates gave excellent support to the dynamic properties of the model.

In order to gain a more precise check on the major period of 3.2 years associated with the model, it seemed desirable to look at the average duration of a large number of business cycles. For this purpose the chronology of U.S. Business Cycles compiled by the National Bureau of Economic Research which extends back to 1854 was used. Arranging the cycles in order or duration, there are 19 cycles under 4 1/2 years and 7 cycles ranging between 6 and 8 years. The longer cycles, which include for example the great depression, are clearly not inventory cycles and were excluded. The average of the three median observations is 3.2 years and the clustering around the median is quite pronounced. This result not only confirms the period of the cycle as given by the model but suggests that essentially the same cyclical mechanism has been active in the economy for a surprisingly long time.

In concluding this paper I would like to reemphasize two points: The first is that a thorough analysis and testing of the dynamic properties of econometric models will increase our ability to discriminate among competing economic hypotheses. There may then result a more orderly growth of tested economic knowledge, with new contributions being additions to, or modifications of, the validated results of earlier workers. The second point is that time series analysis is a natural tool for carrying out a large part of the dynamic testing.

References

1 Goldberger, A.S., Econometric Theory, John Wiley & Sons, New York, 1964

2 Hillinger, C., "An Econometric Model of Mild Business Cycles", The Manchester School, Sept. 1966, pp. 269-284.

3 Johnston, J., Econometric Methods, McGraw-Hill, New York, 1963.

4 Malinvaud, E., Statistical Methods of Econometrics, Rand McNally, Chicago, 1966.

Optimum Control and Synthesis of Organizational Structure of Large Scale Systems

R. Kulikowski

Introduction

It is well known that progress in technical and economic sciences depends much on the solution of complex optimalization problems. In the literature which has been written in recent years on that subject one can discover two basic categories or trends.

The first, which can be called analytic, attempts to decompose or in other words - to break down the complex problem into a number of simpler sub-problems, which can be solved in reasonable time by available computational methods and tools. Then a problem of coordination of sub-problem solutions which renders the integrated problem solution, is being solved. A typical example of that approach is the Dantzig-Wolfe decomposition method in linear programming.

The second trend, which can be called synthetical, assumes that the solutions of a given class of sub-problems are being known and by a process called aggregation attempts to create a complex system with required properties. A typical example of that approach is the so called PERT (Process Evaluation and Review Technique) method which synthesize a minimum-time or minimum-cost project starting with given set of operations having known cost-time relations.

The present paper belongs to the second category.

It is dealing with complex systems including dynamic optimalized production processes, decision making elements (called controllers) and transmission lines which link controllers and processes in a form of hierarchic structure. Many industrial and administrative organizations are organized in that fashion. The higher level controllers send information or resources to the sub-systems which are on the lower levels of the hierarchic structure. The global output produced by sub-systems is considerably reduced in the case of poor transmission or organization of these systems. Therefore the problem of synthesis of organizational structures which are optimal in certain sense represent certain theoretical and practical interest.

In order to solve that problem we shall introduce a measure of quality of organizational structures. Then it will be shown how an optimum structure can be choosen from a given class of possible organizational structures.

2. Optimization and Aggregation

Consider simple organization, shown in Fig. 1, which consists of controlled, dynamic processes, $P_1,\dots,P_n$, local (I-level) controllers $C_1,\dots,C_n$, supervisory (II-level) controller C and transmission lines $L_1,\dots,L_n$ which link C with $C_1,\dots,C_n$.

The operation of controllers is specified by given objective functionals, which together with process equations and constraints can be used for determination of optimum control algorithms. Since in the present paper we are interested mainly in the organizational aspects of complex systems we shall not devote much attention to the derivation of the optimum control algorithms but we shall concentrate on the computation of the so called optimum performance characteristics (O.P.C.) of optimum processes, which are essential for the evaluation of organization quality.

For that purpose let us consider a dynamic process, which is described by a given operator A:

$$y = A(x), \quad y,x \in X,$$

where x is the controlled input, y = output process and X is, generally speaking, a Banach space of functions of time t. Assume that there exist a unique input $\overline{x} \in X$, which minimizes the given objective functional $F(x)$, $x \in X$ (called the control cost), subject to a number of equality or inequality constraints:

$$\phi(x) \geq B,\dots,\Psi(x) \leq Z,$$

where $\phi,\dots,\Psi$ = given functionals in X and B,...,Z = given nonnegative numbers which may represent the desired output production, amount of available resources, optimalized time-interval etc.

Assume that $\overline{x}$ can be effectively derived as a function of time t and B,...,Z. Then it is also possible to derive a function $A = F[\overline{x}(t,B,\dots,Z)] = f[B,\dots,Z]$, which represents the value of control cost as a function of "outer parameters" B,...,Z and which does not depend on time variable t. The function A = f[B,...,Z] will be called O.P.C. of optimalized processes.

Example

Consider a transportation system equipped with electrical motor which should shift an inertial load on the given distance Y, in the given time-interval T with minimum energy consumption.

Assume that the motor is controlled in armature by direct current $x(t) \in L^2[0,T]$. Since the inertial force $m\frac{d^2y}{dt^2}$ (where m is the mass of the load) should be balanced by the motor torque kx(t) (where k = given coefficient) the process eq. becomes: $m\frac{d^2y}{dt^2} = kx(t)$. Then the input-output relation can be written as the integral operator in $L^2[0,T]$:

$$y(t) = y(0) + a\int_0^t (t-\tau)\, x(\tau)\, d\tau, \qquad a = \frac{k}{m}.$$

The optimization problem consists in finding such a function $x(t) \in L^2[0,T]$ which minimizes the energy consumption cost, proportional to

$$F(x) = \int_0^T [x(t)]^2 dt , \tag{1}$$

subject to the constraints:

$$\phi(x) = y(T) - y(0) = a\int_0^T (T-\tau)\, x(\tau)\, d\tau = Y , \tag{2}$$

$$\Psi(x) = \left.\frac{dy(t)}{dt}\right|_{t=T} = a\int_0^T x(\tau)\, d\tau = 0 . \tag{3}$$

The last condition requires that the motor stops at t = T.

In order to solve the present problem we can use variational calculus. The necessary (and in our case also sufficient) condition of optimality requires that the strong gradient of the functional

$$\overline{F}(x) = F(x) + \lambda_1\phi(x) + \lambda_2\Psi(x) ,$$

where λ_1, λ_2 = Lagrange multipliers, is equal zero, i.e.

$$\text{grad}\ \overline{F}(\overline{x}) = \Theta , \tag{4}$$

and

$$\|\text{grad}\ \phi(\overline{x})\| > 0\ ,\ \|\text{grad}\ \Psi(\overline{x})\| > 0$$ *).

From the optimality condition (4) one obtains

$$2\overline{x}(t) + \lambda_1 a(T-\tau) + \lambda_2 a = \Theta\ .$$

Then employing (2), (3) we derive

$$\overline{x}(t) = \frac{3Y}{aT^3}\left(\frac{T}{2} - t\right)\ ,$$

and by (1) we get the O.P.C.

$$A = F(x) = \frac{3}{4}\,\frac{Y^2}{a^2T^2}\ . \qquad (5)$$

Since A is a monotonic function of Y and T it can be easily proved that the same form of O.P.C. one shall obtain when the equality constraint (2) is replaced by $\phi(x) \geq Y$, and when we would like to perform the optimization process in the time not greater than T.

It should be also observed that the same form of O.P.C. one gets when the optimalization should be performed in minimum time subject to the inequality constraints: $F(x) \leq A$, $y(T) - y(0) \geq Y$, or when $y(T) - y(0)$ should be maximalized subject to the constraint $F(x) \leq A$, and when the optimalization time should not exceed T.

In Ref. [1,2,3,4] O.P.C. for many dynamic optimalized processes have been derived. For many cases they assume simple analytic form:

$$A^{\alpha},\ B^{\beta},\ \ldots,\ Y^{\psi},\ Z^{\omega} = (k)^q\ , \qquad (6)$$
$$q = \alpha + \beta + \ldots + \omega$$

where $A,B,\ldots,Y,Z,\alpha,\beta,k$ are positive numbers and $\omega,\psi,\ldots$ negative numbers **). Since the smaller k is the better are properties of the optimalized processes (e.g. in the case of (5): $AT^3Y^{-2} = \frac{3}{4a^2}$ and for fixed T,Y the value of A is small when $k = \frac{\sqrt{3}}{2a}$ is a small number) it can be called the quality index.

Assuming that O.P.C. of subsystems K_i including processes P_i and local

---

*) These conditions follow directly from the well known Lusternik theorem, see e.g. Ref. [5]. Many examples of variational problems in Banach spaces can be found also in Ref. [5].

**) The well known economic model of Cobb-Douglas is a special case of model described by (6).

controllers C_i, $i = 1,2,\ldots,n$ (see Fig. 1) are given: $A_i = f_i(B_i,\ldots,Z_i)$ we can concentrate on the derivation of O.P.C. for aggregated system K, which besides the subsystem K_i consists of supervisory controller C and transmission lines $L_1,\ldots,L_n$.

As the objective function for aggregated system we shall take $\sum_{i=1}^{n} \alpha_i A_i$ and determine such values of $B_i,\ldots,Z_i$ which minimize

$$\sum_{i=1}^{n} \alpha_i A_i = \sum_{i=1}^{n} \alpha_i f_i[B_i,\ldots,Z_i] . \tag{7}$$

subject to the set of aggregated constraints:

$$\sum_{i=1}^{n} \beta_i B_i \leq B , \ldots , \sum_{i=1}^{n} \omega_i Z_i \geq Z , \quad B_i \geq 0 , \ldots , Z_i \geq 0 , \tag{8}$$

where α_i, β_i, $\ldots \geq 1$, ω_i, $\ldots \leq 1$ and $B,\ldots,Z$ are given positive numbers. That represents a nonlinear programming problem. We assume that there exist a unique solution $B_i^0,\ldots,Z_i^0$, $i = 1,2,\ldots,n$ and that it is possible to compute the function

$$A = \sum_{i=1}^{n} \alpha_i f_i[B_i^0,\ldots,Z_i^0] = f[B,\ldots,Z] ,$$

which will be called O.P.C. of aggregated system.

There exist many industrial and economic systems which are aggregated and optimalized according to (7), (8). As an example consider integrated electric power system which consists of n power stations with given performance functions $F_i = f_i(P_i)$, $i = 1,2,\ldots,n$, relating the fuel cost F_i and the amount of power production P_i. The global power production $\sum_{1}^{n} P_i \eta_i$ (where η_i = the so called penelty factors, which represent power losses in transmission lines) should be at least equal to the power demand P and the global fuel cost $\sum_{i=1}^{n} \alpha_i f_i(P_i)$ (where α_i represent fuel losses during transportation) should be minimalized by proper dispatching of power production P_i.

It is possible to show (Ref. [1,2]) that for certain types of O.P.C. the derivation of aggregated O.P.C. is relatively simple. For example in the case of processes having O.P.C. of the form $A_i^\alpha B_i^\beta \ldots Z_i^\omega = (k_i)^q$, $i = 1,\ldots,n$, the aggregated O.P.C. becomes $A^\alpha B^\beta \ldots Z^\omega = (k)^q$, where

$$k = \sum_{i=1}^{n} k_i \lambda_i , \quad \lambda_i = [\alpha_i^\alpha \beta_i^\beta \ldots \omega_i^\omega]^{1/q} , \tag{9}$$

and A_i^o, $B_i^o, \ldots, Z_i^o$ can be determined from linear equations. A similar property have the functions

$$A_i = \beta_i f \left[\frac{B + \bar{B}_i}{\beta_i} \right] + a_i \tag{10}$$

where β_i, a_i, $\bar{B}_i$ given numbers and f is monotonic differentiable function having unique inverse $[f']^{-1}$ (ref. [3]). It is also possible to show that a continuous O.P.C. can be piecewise approximated by functions (10) with the desired degree of accuracy.

It should be observed that when O.P.C. are described by function of type (9) or (10) the aggregation and optimalization processes can be applied to multilevel structures yielding at each stage the same form of O.P.C. with quality indexes which can be derived by simple relations of the type (9). In other words the amount of variables or information which comes into account at each control level is strictly limited.

It is also possible to evaluate qualities of different organizational structures. Assume for example that three different processes, described by (6), with performances k_1, k_2, k_3 and three different organizations shown in Fig. 2a,b,c, are given. The corresponding quality indexes, derived by (9), become

$$k_a = \lambda_1 k_1 + \lambda_2 k_2 + \lambda_3 k_3 , \tag{11}$$

$$k_b = \lambda_1 k_1 + \lambda_{23} \lambda_2 k_2 + \lambda_{23} \lambda_3 k_3 , \tag{12}$$

$$k_c = \lambda_{12} \lambda_1 k_1 + \lambda_{12} \lambda_2 k_2 + \lambda_3 k_3 , \tag{13}$$

where $\lambda_1, \lambda_2, \lambda_3, \lambda_{12}, \lambda_{23}$ represent losses introduced by transmission lines which link the respective controllers. Now we are able to compare different organizations what will be done in the next section.

3. Synthesis and Optimum Control of Organizational Structures

As follows from (9) (compare also (11), (12), (13)) the resulting quality index of an organization which consists of n controlled processes with given performance indexes $k_1, k_2, \ldots, k_n$ can be written in the form

$$k = \sum_{i=1}^{n} k_i l_i , \tag{14}$$

where l_i = loss indices depending on organization structure. It is also obvious that the smaller k is the better is the global system performance.

The minimum value of k can be obtained by:

(a) assigning processes to the given fixed structure, i.e. to the given, ordered set $\{l_i\}_o^n$:

$$l_1 \leq l_2 \leq \ldots \leq l_n ,$$

one should assign indexes ν in the set $\{k_\nu\}^n$ in such a way that (14) is minimum;

(b) allowed reorganization of structure, by changing the position of controllers and transmission lines, which decrease the value of (14).

As far as the assignment problem is concerned the following two theorems may represent certain interest.

Theorem I.

Let two sets $\{l_\nu\}_1^n$, $\{k_j\}_1^n$ of positive numbers be given. The set of $K=\sum_1^n k_i l_i$, corresponding to any possible assignment of indexes ν, j, is contained in the interval

$$\left[\frac{2\overline{l}\overline{k}}{\sqrt{\frac{lK}{Lk}}+\sqrt{\frac{Lk}{lK}}} , \overline{l}\overline{k} \right] \tag{15}$$

where

$$\overline{l} = \{ \sum_{i=1}^n l_i^2\}^{1/2} , \quad \overline{k} = \{ \sum_{i=1}^n k_i^2\}^{1/2}$$

$$l = \min_i l_i , \quad k = \min_i k_i , \quad L = \max_i l_i , \quad K = \max_i k_i .$$

The upper bound ($\overline{l}\overline{k}$) is atteined if and only if $k_i = \alpha l_i$, $i = 1,\ldots,n$, α = const. The lower bound in (15) is attained if and only if $p = \frac{L/l}{L/l + K/k}$ is an integer and

$$\begin{matrix} k_i = k \\ l_i = L \end{matrix} \quad i = 1,2,\ldots,p , \qquad \begin{matrix} k_i = K \\ l_i = l \end{matrix} \quad i = p+1,\ldots,n .$$

Proof of that theorem is based on the known Cauchy and G. Polya and C. Szego inequalities [4].

Theorem II.

Let the sets $\{l\}_1^n$, $\{k\}_1^n$ of positive numbers be given. The value of $K = \sum_1^n l_i k_i$ is minimum if

$$k_1 \leq k_2 \leq \dots \leq k_n , \quad l_1 \geq l_2 \geq \dots \geq l_n , \tag{16}$$

or if

$$k_1 \geq k_2 \geq \dots \geq k_n , \quad l_1 \leq l_2 \leq \dots \leq l_n . \tag{17}$$

These conditions become also necessary in the case of strict inequalities in (16), (17).

The validity of that theorem for $n = 2$ is obvious. For $n > 2$ if can be proved by induction (see also [4]).

Example

Consider two organizations shown in Fig. 2b,c, and assume that $\lambda_1 = \lambda_2 = \lambda_3 = \lambda_{12} = \lambda_{23} = \lambda > 1$.

In the case of system 2b we have $l_1 = \lambda < l_2 = \lambda^2$ and $l_2 = l_3 = \lambda^2$. Then according to theorem II that organization is optimum if $k_1 \geq k_2 \geq k_3$. For the same reason organization shown in Fig. 2c becomes optimum if $k_1 \leq k_2 \leq k_3$. These structures become equivalent when $\lambda = 1$.

The last example indicates that besides optimum control of processes in order to get best results it is necessary to reorganize the structure when quality indexes of the subsystems change in time. In other words the higher level controllers should reorganize the system structure if necessary.

The theorem II can be used also for synthesis of multilevel structures. As an example assume that n controllers C_i, $i = 1,2,\dots,n$, and N processes, equipped with local controllers, so they can be completely described by indexes $k_1 \leq k_2 \leq \dots \leq k_N$, be given. Transmission losses are assumed the same for each interconnection and equal $\lambda > 1$. The maximum amount of processes which can be controlled by controllers C_i are equal m_i , $i = 1,2,\dots,n$, respectively and

$$m_1 \leq m_2 \leq \dots \leq m_n , \quad \sum_{i=1}^n m_i = N .$$

Besides we assume that each controller can also optimize one subsystem of controllers and processes. The problem consists in determining the best organization of controllers and processes.

Let us observe that the numbers

$$K_i = \sum_{j=m_i'}^{m_i} k_j , \quad m_i' = m_{i-1} + 1, \quad i = 1,2,\ldots,n$$

satisfy the condition: $K_1 \leq K_2 \leq \ldots \leq K_n$, and the possible organizations will give loss-coefficients of the form λ^k, $k = 1,2,\ldots,n$.

Then using theorem II one can obtain the structure shown in Fig. 3, having the quality index

$$K = \sum_{i=1}^{n} K_i l_i , \tag{18}$$

where

$$l_i = \lambda^{n-i+1} , \quad \text{and} \quad l_1 > l_2 > \ldots > l_n > 1 .$$

This organization is optimum in the sense that no allowed reorganization (consisting in exchanging processes and subsystems) exists which would decrease the value of K given by (18).

Other examples of synthesis of organizations and some extensions are given in Ref. [4]. For example one can assume that the numbers of processes m_i are not fixed and the loss coefficients k_i are increasing functions of m_i. In that case the optimum number of control-levels depends, generally speaking, on the global number of processes N.

As an example assume $K_i = k$, $\lambda_i = \lambda$, $i = 1,2,\ldots,N$, $N = m^2 = \text{const.}$, and compare resulting losses for the single-level (l_I) and two-levels (l_{II}) organizations of the type shown in Fig. 2a and Fig. 2b,c.

One obtains

$$l_I = m^2 \cdot \lambda(m^2), \quad l_{II} = m^2[\lambda(m)]^2 .$$

When m increases there exists, generally speaking, such a number $m = m_o$, that $l_{II} < l_I$. Assume e.g. $\lambda(m) = 1 + \delta m$. Then

$$l_I(m) = m^2(1 + \delta m^2) > l_{II}(m) = m^2[1 + \delta m]^2$$

when

$$m > \frac{2}{1 - \delta} .$$

4. Multi-level Optimization of Stochastic, Interacting Systems

So far we were considering optimization of deterministic, independent dynamic processes by local (I level) and supervisory (II or higher level) controllers. The goal of I-level controller was to optimize the dynamic processes subject to the given amount of control resources which were delivered by II-level, whereas the II-level controller distributed the received resources among the I-level subsystems. In the systems of that type we did not need feedback loops between processes and I-level controller or between I and II level controllers. In the more complicated situation when the O.P.C. of subsystems are interacting and it is required to preserve the decentralized control by I and II level controllers feedback loops are, generally speaking, necessary (because they provide an exchange of information between I and II level).

In order to explain that problem we shall start with a simple stabilization system which consists of n subsystems including processes P_i and local controllers C_i, and which is shown in Fig. 4 for n = 2. The input-output relations for the subsystems $y_i = f_i[x_i]$, and the additive interactions $z_{ji} = \varphi_{ji}(x_j)$ are given continuous functions, $i,j = 1,2,\ldots,n$. It is desired to obtain the resulting outputs

$$f_i[x_i] + \sum_{\substack{j=1 \\ j\neq i}}^{n} \varphi_{ji}(x_j) \ , \quad i = 1,2,\ldots,n$$

equal to the given numbers Y_i.

If the interactions were not present each controller C_i could determine the required control values $x_i = \overline{x}_i$ by solving the eqs.

$$f_i[x_i] = Y_i \ , \quad i = 1,2,\ldots,n \ .$$

For that purpose it is convenient to solve the equivalent eqs.:

$$x_i = x_i + a_i[Y_i - f_i(x_i)] = F_i(x_i) \ , \quad i = 1,\ldots,n \tag{19}$$

where the numbers a_i are choosen in such a way that the functions $F_i(x_i)$ satisfy the contraction conditions in the intervals X_i including x_i:

$$|F_i(x_i') - F_i(x_i'')| \leq \alpha|x_i' - x_i''| \ , \quad \alpha < 1, \quad i = 1,2,\ldots,n \ ,$$

x_i', x_i'' = arbitrary points in X_i.

Then the values x_i can be derived by iterations

$$x_i^{(k+1)} = F_i(x_i^{(k)}), \quad k = 0,1,2,\dots,n, \quad i = 1,2,\dots,n, \tag{20}$$

starting with the arbitrary values $x_i^{(o)} \in X_i$, $i = 1,2,\dots,n$.

It is well known that $\lim_{k\to\infty} x_i^{(k)} \longrightarrow \overline{x}_i$ and the obtained solution is unique.

The iterations can be also used when the explicit form of input-output relations is unknown but the controllers can observe the outputs $y_i^{(k)}$, which correspond to the fixed input $x_i^{(k)}$, using feedback loops (denoted by dotted line in Fig. 4). Since those observations are frequently influenced by random noise one is interested in present case in the expected values of $y_i^{(k)}(\omega)$, i.e.

$$E\{y_i^{(k)}(\omega)|x_i^{(k)}\} = f_i(x_i^{(k)}), \quad i = 1,2,\dots,n \quad k = 0,1,\dots$$

where ω is random variable.

The functions $f_i(x_i)$ should be now treated as regression functions and the problem which faces us is the solution of regression eqs.:

$$f_i[x_i] - Y_i = 0, \quad i = 1,2,\dots,n,$$

by iterations, using values $y_i^{(k)}(\omega)$ taken from observations.

That can be done by the so called stochastic approximations, having the form:

$$x_i^{(k+1)}(\omega) = x_i^{(k)}(\omega) + a_n[Y_i - y_i^{(k)}(\omega)], \quad k = 0,1,\dots, \quad i = 1,\dots,n,$$

which, as shown by Robbins, Monro [6] will converge stochastically to the values $\overline{x}_i$, $i = 1,\dots,n$, i.e.:

$$\lim_{k\to\infty} E\{|x_i^{(k)}(0) - \overline{x}_i|\} = 0$$

if certain regularity conditions hold. The regularity conditions, besides contractions, include the requirement that numbers C_1, C_2, C_3 exist such that

1. the probability $P\{|y(x)| < C_1\} = 1$
2. $\dfrac{C_2}{k} \leq a_{k-1} \leq \dfrac{C_3}{k}$, $k = 1,2,\dots,$

and the dispersion of $x_i^{(o)}$ is finite.

The iterations can be also used in the case when interactions are present. In that case instead of (19), (20) we get

$$x_i^{(k+1)} = x_i^{(k)} + a_i[Y_i' - f_i(x_i^{(k)})] = F_i(x_1^{(k)},\dots,x_n^{(k)}) \,, \qquad (21)$$

$$Y_i' = Y_i - \sum_{\substack{j=1 \\ j\neq i}}^{n} \varphi_{ji}(x_j^{(k)}) \,, \qquad (22)$$

or using vector notation (21), (22) can be written:

$$x^{(k+1)} = F[x^{(k)}] \,, \quad k = 0,1,\dots, \qquad (23)$$

where $x \equiv [x_1,x_2,\dots,x_n]$, $F \equiv [F_1,F_2,\dots,F_n]$ is a nonlinear continuous operator in n dimensional space E^n. When F is a contracting operator in a set $X \subset E^n$, and $x^0 \in X$, iterations (23) will converge to the unique solution $\overline{x} \equiv [\overline{x}_1, \overline{x}_2,\dots,\overline{x}_n] \in X$.

The calculations corresponding to (21), (22) can be implemented in the two-levels form, shown in Fig. 4 for n = 2, where the II-level controller C derives the values Y_I by (22) and the I-level controllers derive $x_i^{(k+1)}$ by (21). The advantage of two-level process is that it utilizes the same control algorithms for I level as in the case without interactions. However, it requires the exchange of information between I and II level controllers.

The control processes (21), (22) can be also realized in the case when instead of $f_i[x_i^{(k)}]$, $\varphi_{ji}[x_j^{(k)}]$ the values $y_i^{(k)}(\omega)$, $z_{ji}^{(k)}(\omega)$, obtained by observations, are being used. In that case, instead of (21), (22), we get the following algorithms

for I level

$$x_i^{(k+1)} = x_i^{(k)} + a_i^{(k)}[Y_i'^{(k)}(\omega) - y_i^{(k)}(\omega)], \qquad (24)$$

$$i = 1,2,\dots,n, \quad k = 0,1,\dots,$$

for II level

$$Y_i'^{(k)}(\omega) = Y_i - \sum_{\substack{j=1 \\ j\neq i}}^{n} z_{ji}^{(k)}(\omega) \,, \quad i = 1,2,\dots,n \,. \qquad (25)$$

If the regularity conditions for the multidimensional case hold (see in that respect Ref. [7]) the two level iteration process (24), (25) converges stochastically to the solutions $\overline{x}_i$, $i = 1,\dots,n$, of the regression eqs.:

$$Y_i - f_i(x_i) - \sum_{\substack{j=1 \\ j\neq i}}^{n} \varphi_{ji}(x_j) = 0 \,.$$

One should observe that the stochastic approximations can be also used for decomposition of complex optimization problems.

Consider, for example, the problem of finding the optimum values x_i^o, $i = 1,\dots,n$, which maximize $y = \sum_1^n y_i$ in the aggregated system which consists of n subsystems with O.P.C. $y_i = F_i(x_i)$, where F_i are differentiable functions, subject to the aggregation interactions $\sum_{i=1}^n x_i = x$, where x = given positive number.

If the interactions were not present and the functions $\varphi_i(x_i) = x_i + a_i F_i'(x_i)$ (where a_i = given numbers) were contracting in X_i we could derive the unique solutions

$$x_i^o = \lim_{k \to \infty} x_i^{(k)} , \text{ where } x_i^{(k+1)} = \varphi_i[x_i^{(k)}] , \quad i = 1,2,\dots,n$$

starting with arbitrary values $x_i^{(o)} \in X_i$. It is also possible to use stochastic approximations of the Kiefer, Wolfowitz [8] type:

$$x_i^{(k+1)}(\omega) = x_i^{(k)}(\omega) + \frac{a_i^{(k)}}{c_i^{(k)}} \left[\bar{\bar{y}}_i^{(k)}(\omega) - \bar{y}_i^{(k)}(\omega) \right], \qquad (24)$$

$$i = 1,\dots,n , \quad k = 0,1,\dots,$$

where

$\bar{y}_i^{(k)}(\omega)$ is the subsystem "i" output when $x_i^{(k)}(\omega) - c_i^{(k)}$ is the input,

$\bar{\bar{y}}_i^{(k)}(\omega)$ is the subsystem "i" output when $x_i^{(k)}(\omega) + c_i^{(k)}$ is the input,

$c_i^{(k)}$ = given numbers.

It is well known [8] that under certain regularity conditions, regarding the repression functions $F_i(x_i)$ and

$$c_i^{(k)} \longrightarrow 0, \quad \sum_{k=0}^{\infty} a_i^{(k)} = \infty , \quad \sum_{k=0}^{\infty} a_i^{(k)} c_i^{(k)} < \infty , \quad \sum_{k=1}^{\infty} \left(\frac{a_i^{(k)}}{c_i^{(k)}} \right)^2 < \infty$$

$$i = 1,2,\dots,n$$

the processes (24) converge stochastically to x_i^o.

In the case when interaction $\sum_{i=1}^n x_i = X$ is present it is possible to construct two-level optimization algorithm:

I level:

$$x_i^{(k+1)}(\omega) = x_i^{(k)}(\omega) + \frac{a_i^{(k)}}{c_i^{(k)}} \left[\bar{\bar{y}}_i^{(k)}(\omega) - \bar{y}_i^{(k)}(\omega) \right] + b_i^{(k)} z^{(k)}$$

$$k = 0,1,2,\dots, \quad i = 1,\dots,n ,$$

II level:

$$Z^{(k)}(\omega) = X - \sum_{i=1}^{n} x_i^{(k)}(\omega) .$$

That process will converge stochastically to the solution x_i^o, starting with arbitrary $x_i^{(o)}$ when corresponding contraction and regularity conditions hold.

It is possible to construct many different examples of optimization of non-linear, interacting systems by stochastic approximations and implement the solution in a multi-level, decentralized system. The advantage of such an approach is that it uses simple computation algorithms. However, as far as the convergence speed is concerned this method is not most effective (in the statistical sense). Using methods of decision functions theory one can obtain more effective results but the computational algorithms are, as a rule, more complicated.

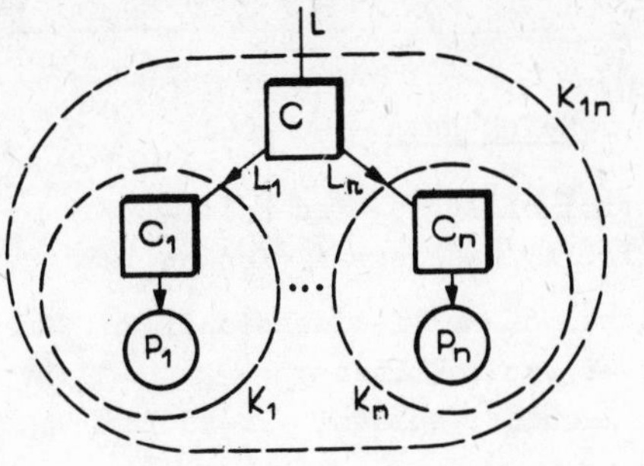

Fig.1.

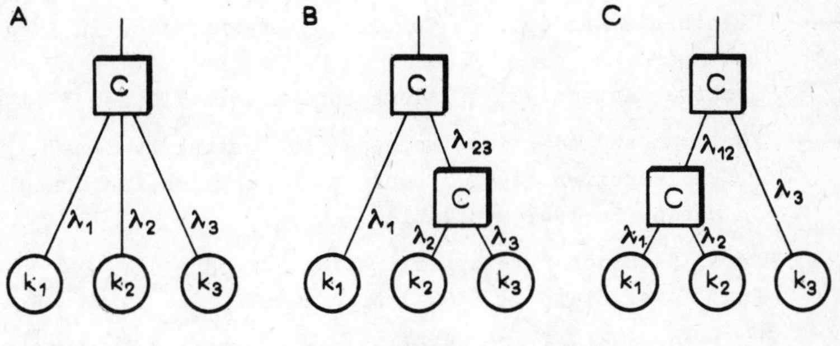

Fig. 2.

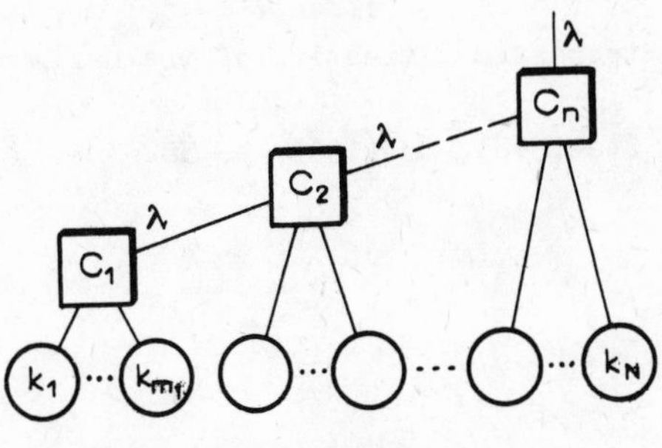

Fig. 3.

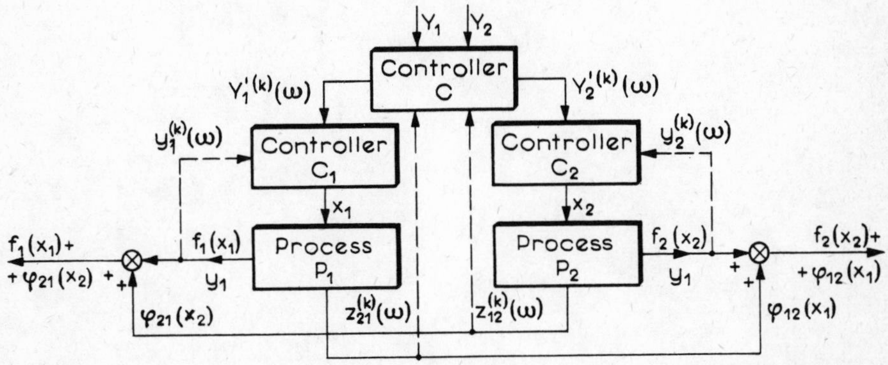

Fig.4.

References

1. R. Kulikowski: "Optimum Control of Aggregated Multi-level Systems
Proc. of III IFAC Congress in London 1966

2. ———————— "Optimum Control of Multi-dimensional and Multi-level Systems" in "Advances of Control Theory"
Edited by C. Leondes, Academic Press Inc. New York (1966)

3. ———————— "Optimalization of Aggregated Dynamic Systems"
Archiwum Automatiyki i Telemechaniki vol. XI, No. 3 (1966)

4. ———————— "Synthesis and Optimum Control of Organization in Large Scale Systems"
Archiwum Automatyki i Telemechaniki vol. XII No. 3 (1967)

5. ———————— "Optimum and Adaptive Processes in Control Systems"
P.W.N. Warszawa (1965) (book in Polish, Russian translation published in Moscow in 1967 available)

6. H. Robbins, S. Monro: "A Stochastic Approximation Method"
Ann. Math. Stat. 22 (1951) p. 400-407

7. J.R. Blum: "Multidimensional Stochastic Approximation Methods"
Ann. Math. Statistics 25 (1954) p. 737-744

8. J. Kiefer, J. Wolfowitz: "Stochastic Estimation of the Maximum of a Regression Function"
Ann. Math. Stat. vol. 25 (1954) p. 382-386

OPTIMAL ACCUMULATION IN A LISTIAN MODEL

Harl E. Ryder, Jr.

The most respectable of the traditional arguments against the classical doctrine of free trade has been the "infant industry" argument of Hamilton[1] and List[2], which appeals to dynamic considerations ignored in the classical comparative statics analysis. It is natural to re-examine this controversy in the light of recent advances in the dynamic analysis of economic growth. In another essay[3] I examined a model of optimal patterns of investment and foreign trade in a two-sector growth model. In that model, however, there was no provision for an "infant industry"; hence, the restrictions on trade resulted not from dynamic considerations of optimal growth, but rather from essentially static considerations of monopoly power. The purpose of this paper is to develop an optimal growth model which exhibits the essential feature of the Listian model--the effect of foreign trade on endogenous technical change.

The model is defined and justified in the first section. In section 2 we construct paths satisfying the necessary conditions for an optimum and indicate how to choose among them when there are several. Section 3 traces the development process for a single country facing a constant world price. Section 4 does the same for a multicountry world in which each country behaves optimally in anticipation of a constant world price, while that price is determined by their aggregate behavior.

---

1. Alexander Hamilton, Report on the Subject of Manufactures, 1791; reprinted in The Papers of Alexander Hamilton, Vol X, Harold C. Syrett, ed., Columbia University Press, New York, 1966, pp. 230-340.

2. Friedrich List, Das Nationale System der Politischen Oekonomie, 1844.

3. Harl E. Ryder, Jr., "Optimal Accumulation and Trade in an Open Economy of Moderate Size", Essays on the Theory of Optimal Economic Growth, Karl Shell, ed., M.I.T. Press, Cambridge, Massachusetts, 1967, pp. 87-116.

1. Definitions and Assumptions

The basic idea of the "infant industry" argument seems to be that the productivity of a given industry will be higher after it has become well established than it was when first engaged in. In particular, this effect is thought to be more pronounced in manufacturing than in agriculture. Hamilton claims, for example, "that the establishment and diffusion of manufactures have the effect of rendering the total mass of productive labor in a community, greater than it would otherwise be"[4].

We shall consider a two-sector economy producing an investment good and a consumption good by means of two factors, capital and labor. We shall assume that both sectors are equally capital intensive, but that the List effect occurs only in the investment good sector. The economy can engage in balanced trade on fixed terms. The productivity of the investment good sector depends on the amount of recent experience in that sector.

More precisely, we have:

$$Y_C = F(K_C, L_C) \tag{1}$$

$$Y_I = A(G)\, F(K_I, L_I) \tag{2}$$

$$K = K_I + K_C \tag{3}$$

$$L = L_I + L_C \tag{4}$$

$$X_C = Y_C + Z_C \tag{5}$$

$$X_I = Y_I + Z_I \tag{6}$$

$$0 = p\, Z_I + Z_C \tag{7}$$

$$\dot{L} = n\, L \tag{8}$$

$$\dot{K} = X_I - \mu K \tag{9}$$

$$\dot{G} = Y_I - vG \tag{10}$$

---

4. Op. cit., p. 246

where

Y_j is the output of the j sector (j = I,C)

K_j is the capital employed in the j sector (j = I,C)

L_j is the labor employed in the j sector (j = I,C)

G is the amount of recent experience in the I sector

X_j is the absorption of the j good (j = I,C)

Z_j is the amount of the j good imported (j = I,C)

p is the international price of I in terms of C

n is the rate of growth of the labor force

μ is the rate of depreciation of capital

v is the rate of deterioration of experience

The production function F is assumed to satisfy the usual neoclassical conditions:

$$
(11)\quad \begin{cases} F(K,L) > 0,\ F(0,L) = F(K,0) = 0,\ F(\theta K,\theta L) = \theta F(K,L) \\ \frac{\partial F}{\partial K} > 0,\ \frac{\partial F}{\partial L} > 0,\ \frac{\partial^2 F}{\partial K \partial L} > 0 \\ \frac{\partial}{\partial K} F(0,L) = \frac{\partial}{\partial L} F(K,0) = \infty \end{cases}
$$

The productivity function A is assumed to satisfy:

$$
(12)\quad \begin{cases} A(G) > 0,\ A'(G) > 0,\ A''(G) < 0. \\ A(0) = \underline{A} > 0,\ A(\infty) = \overline{A} \end{cases}
$$

We have endogenous technical change of a sort that might be described as "learning by doing", although it differs from the concept introduced under that name by Arrow[5]. In Arrow's model, "learning by doing" means learning how to make more efficient investment goods; in this model, it means learning how to make investment goods more efficiently. Technical progress of this sort might develop from the mastery of the skills required by this industry, reduction of spoiled materials, R & D directed against the major bottlenecks at any given stage of technical knowledge, etc. By the form of equation (2) the fruits of experience are accessible to all producers. Thus, experience is a measure of what Meade calls "atmosphere - creating external economies"[6].

---

5. Kenneth J. Arrow, "The Economic Implications of Learning by Doing", Review of Economics Studies 29 (June, 1962), 155-173.

6. J.E. Meade, Trade and Welfare, Oxford University Press, London, 1955, p. 256.

The condition $A''(G) < 0$ means that such progress is easiest when the level of experience is low and becomes progressively more difficult when the most obvious improvements have already been made. The condition $A(\infty) = \overline{A}$ means that the productivity effect of experience is bounded. The condition $A(0) = \underline{A}$ means that a positive output is possible even without experience.

To simplifiy the analysis, let us assume that $n = 0$, $L = 1$. With both sectors equally capital intensive, efficiency requires that

$$\frac{K_I}{K} = L_I = s, \quad \frac{K_C}{K} = L_C = 1 - s.$$

Then, since F is homogeneous, we can define $f(K) = F(K,1)$. Now equations (1) - (10) reduce to

(13) $X_C = (1 - s) f(K) + Z_C$

(14) $\dot{K} = s A(G) f(K) - \frac{Z_C}{p} - \mu K$

(15) $\dot{G} = s A(G) f(K) - vG$

(16) $0 \leq s \leq 1$

(17) $-(1 - s) f(K) \leq Z_C \leq p\, s\, A(G)\, f(K)$

where

(18) $\begin{cases} f(K) > 0,\ f'(K) > 0,\ f''(K) < 0 \\ f(0) = f'(\infty) = 0,\ f'(0) = f(\infty) = \infty. \end{cases}$

We wish to maximize

(19) $V = \int_0^\infty X_C e^{-\delta t} dt$

subject to (13) - (17) and

(20) $K(0) = K_0$, $G(0) = G_0$.

Let us introduce auxiliary variables q, γ the shadow prices of capital and experience respectively. We define

(21) $H = (1 - s) f(K) + Z_C + q \left[sA(G)f(K) - \frac{Z_C}{p} - \mu K\right] + \gamma \left[sA(G)f(K) - vG\right].$

For an optimal solution there must exist q, γ satisfying:

(22) the s and Z_C at every point in time maximize H subject to (16), (17).

(23) $\dot{q} = \delta q - \left.\frac{\partial H}{\partial K}\right|_{(22)}$

(24) $\dot{\gamma} = \delta\gamma - \left.\frac{\partial H}{\partial G}\right|_{(22)}$

(25) $\lim_{t \to \infty} qe^{-\delta t} = 0, \lim_{t \to \infty} \gamma e^{-\delta t} = 0.$

The economic interpretation of these conditions is clear. H as defined by (21) is the value of net output, where q is the value of a unit of investment (installation of the capital good) and γ is the value of a unit of experience (production of the capital good). Condition (22) means that those goods are produced and traded which maximize the value of output at each instant. Equations (23) and (24) mean that the sum of the marginal product and the capital gain on a unit of capital or experience is equal to interest at rate δ . Condition (25) means that no value is to be attributed to capital or experience at the end of the infinite period.

Decentralized decisions are possible in this model. If the planner establishes a price q for the capital good and a subsidy γ for domestic production of the capital good, then competitive behavior between firms and the owners of capital and labor will lead to the proper allocation of resources. Experience cannot be left entirely to the market for two reasons. Since it is an external economy, any firm can use it without paying a rent or royalty to the firm that produced it. Even if it were possible to require such payments, it would not be efficient to do so. Since the production function is homogeneous of degree one in capital and labor with experience held constant, wages and rent on capital will exactly exhaust the product. If firms also had to pay the marginal product of experience, their profits would be negative.

2. Construction of Optimal Paths

We can divide the phase space of (K, G, q, γ) into four regions to which correspond the four corner solutions of (22). We shall designate these patterns using Roman numerals for the pattern of production (I--specialized in the consumption good; III--specialized in the investment good) and capital letters for the pattern of absorption (A--specialized in consumption; C--specialized in investment). Nonspecialization will be designated by II and B for production and absorption respectively.

When q < p we are in pattern A. Z_C takes its maximum value, $Z_C = p\ s\ A(G)\ f(K)$. Substituting into (21) gives

(21a) $H = \left[(1 - s) + s(p + \gamma)\ A(G)\right] f(K) - q\,\mu K - \gamma\nu G.$

Therefore, when $(p + \gamma) A(G) - 1 \gtrless 0$ we will be in patterns IIIA, IIA, or IA, respectively.

When $q > p$ we are in pattern C. Z_C takes its minimum value, $Z_C = -(1-s) f(K)$. Substituting into (21) gives

$$(21c) \quad H = \left[(1 - s) \frac{q}{p} + s(q + \gamma) A(G)\right] f(K) - q \mu K - \gamma v G.$$

Therefore, when $(q + \gamma) A(G) - \frac{q}{p} \gtrless 0$ we will be in patterns IIIC, IIC, or IC, respectively.

Table I gives the region corresponding to each pattern and the values which s and Z_C must assume. Table II summarizes differential equations in each pattern.

The motion of K and G in each of the four corner patterns is shown in Figures 1 - 4. The motion of K and G in the remaining five patterns will be a linear combination of the motion in the four corner patterns.

Table I

	I	II	III
A	$q < p$ $\gamma < \frac{1}{A(G)} - p$ $s = 0$ $Z_C = 0$	$q < p$ $\gamma = \frac{1}{A(G)} - p$ $0 \leq s \leq 1$ $Z_C = p s A(G) f(K)$	$q < p$ $\gamma > \frac{1}{A(G)} - p$ $s = 1$ $Z_C = p A(G) f(K)$
B	$q = p$ $\gamma < \frac{1}{A(G)} - p$ $s = 0$ $-f(K) \leq Z_C \leq 0$	$q = p$ $\gamma = \frac{1}{A(G)} - p$ $0 \leq s \leq 1$ $-(1-s) f(K) \leq Z_C \leq p s A(G) f(K)$	$q = p$ $\gamma > \frac{1}{A(G)} - p$ $s = 1$ $0 \leq Z_C \leq p A(G) f(K)$
C	$q > p$ $\gamma < q(\frac{1}{pA(G)} - 1)$ $s = 0$ $Z_C = -f(K)$	$q > p$ $\gamma = q(\frac{1}{pA(G)} - 1)$ $0 \leq s \leq 1$ $Z_C = -(1-s) f(K)$	$q > p$ $\gamma > q(\frac{1}{pA(G)} - 1)$ $s = 1$ $Z_C = 0$

Table II

	I	II	III
A	$\dot{K} = -\mu K$ $\dot{G} = -vG$ $\dot{q} = (\mu + \delta)\ q - f'(K)$ $\dot{\gamma} = (v + \delta)\gamma$	$\dot{K} = -\mu K$ $\dot{G} = sA(G)\ f(K) - vG$ $\dot{q} = (\mu + \delta)\ q - f'(K)$ $\dot{\gamma} = (v + \delta)\gamma - sA'(G)f(K)(p + \gamma)$	$\dot{K} = -\mu K$ $\dot{G} = A(G)\ f(K) - vG$ $\dot{q} = (\mu + \delta)q - A(G)f'(K)(p + \gamma)$ $\dot{\gamma} = (v + \delta)\gamma - A'(G)f(K)(p + \gamma)$
B	$\dot{K} = -\frac{Z_C}{p} - \mu K$ $\dot{G} = -vG$ $\dot{q} = (\mu + \delta)\ q - f'(K)$ $\dot{\gamma} = (v + \delta)\gamma$	$\dot{K} = sA(G)\ f(K) - \frac{Z_C}{p} - \mu K$ $\dot{G} = sA(G)\ f(K) - vG$ $\dot{q} = (\mu + \delta)\ q - f'(K)$ $\dot{\gamma} = (v + \delta)\gamma - sA'(G)f(K)(q + \gamma)$	$\dot{K} = A(G)\ f(K) - \frac{Z_C}{p} - \mu K$ $\dot{G} = A(G)\ f(K) - vG$ $\dot{q} = (\mu + \delta)q - A(G)f'(K)(p + \gamma)$ $\dot{\gamma} = (v + \delta)\gamma - A'(G)f(K)(p + \gamma)$
C	$\dot{K} = \frac{f(K)}{p} - \mu K$ $\dot{G} = -vG$ $\dot{q} = [\mu + \delta - \frac{f'(K)}{p}]\ q$ $\dot{\gamma} = (v + \delta)\gamma$	$\dot{K} = f(K)[\frac{1-s}{p} + sA(G)] - \mu K$ $\dot{G} = sA(G)\ f(K) - vG$ $\dot{q} = [\mu + \delta - \frac{f'(K)}{p}]\ q$ $\dot{\gamma} = (v + \delta)\gamma - sA'(G)f(K)(q + \gamma)$	$\dot{K} = A(G)\ f(K) - \mu K$ $\dot{G} = A(G)\ f(K) - vG$ $\dot{q} = (\mu + \delta)q - A(G)f'(K)(q + \gamma)$ $\dot{\gamma} = (v + \delta)\gamma - A'(G)f(K)(q + \gamma)$

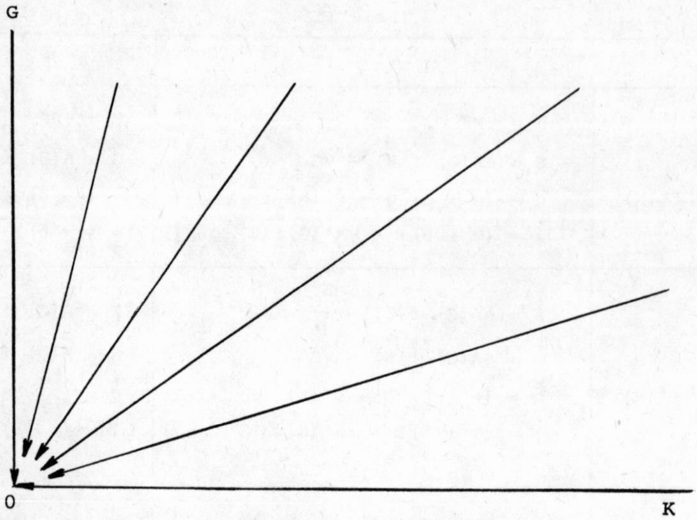

Figure 1
Motion of K and G in Pattern IA

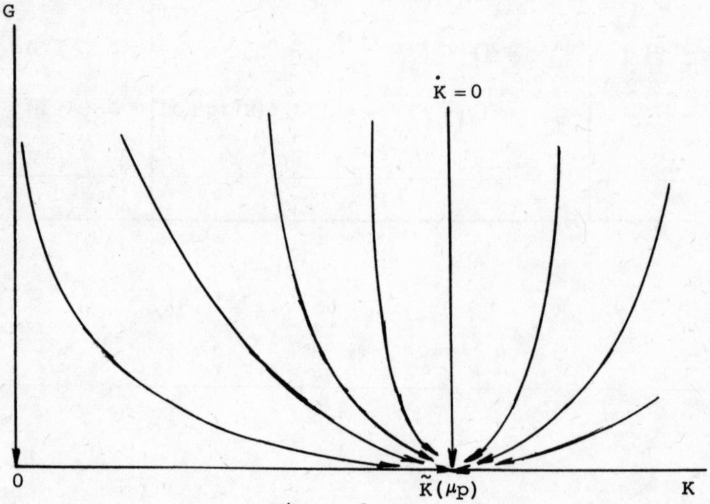

Figure 2
Motion of K and G in Pattern IC

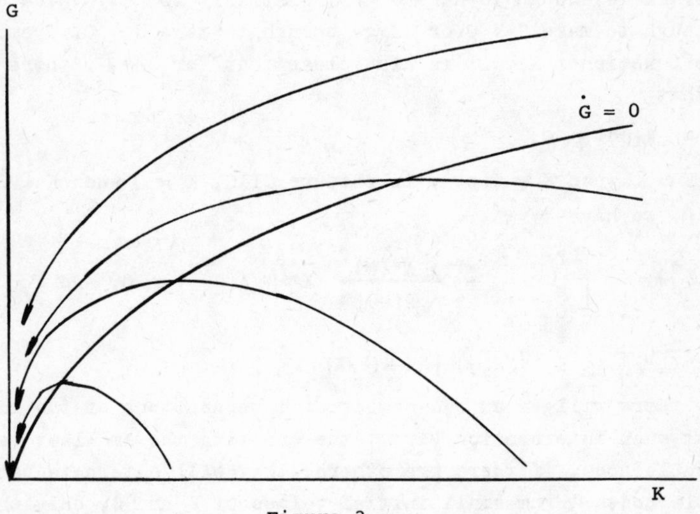

Figure 3
Motion of K and G in Pattern IIIA

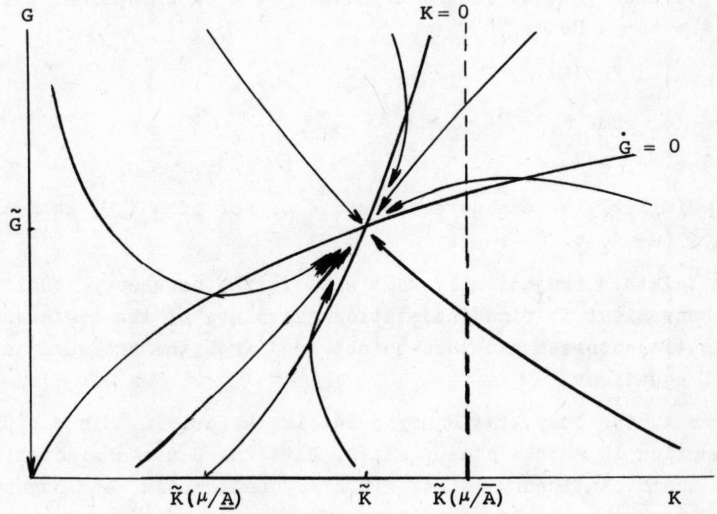

Figure 4
Motion of K and G in Pattern IIIC

In patterns I, G is always falling; in patterns A, K is always falling. In patterns III, $\dot{G} = 0$ when

(26) $f(K) = \frac{vG}{A(G)}$. This locus is an upward sloping curve which passes through the origin. It is easy to verify that for any value of K, G can be chosen small enough to make $\dot{G} > 0$ or large enough to make $\dot{G} < 0$. Similarly for any value of G, K can be chosen small enough to make $\dot{G} < 0$ or large enough to make $\dot{G} > 0$. From the assumptions (18) on the function f(K), it is also clear that for any μ , there is a unique $\tilde{K}(\mu) > 0$ such that

(27) $f[\tilde{K}(\mu)] - \tilde{K}(\mu) = 0.$

In pattern IC, $\dot{K} = 0$ when $K = \tilde{K}(p\mu)$; in pattern IIIC, $\dot{K} = 0$ when $K = \tilde{K}[\mu/A(G)]$. Since $\tilde{K}'(\mu) < 0$, we have

$$(28)\quad \begin{cases} \left.\frac{dK}{dG}\right|_{\dot{K}=0} = -\frac{\mu A'(G)}{[A(G)]^2}\tilde{K}'[\mu/A(G)] > 0, \text{ and} \\ \tilde{K}(\mu/\underline{A}) \leq \tilde{K}[\mu/A(G)] < \tilde{K}(\mu/\overline{A}). \end{cases}$$

In pattern IIIC there will be an odd number of intersections of the loci $\dot{G} = 0$ and $\dot{K} = 0$. The first such intersection (i.e., the one with the smallest values of K and G) will be a stable node. If there are others, they will alternate between saddle points and stable nodes. From small initial values of K and G, only the first one will be accessible. Call it $(\tilde{K}, \tilde{G})$.

It is easy to show that we can never have $\gamma < 0$ on an optimal path. In all patterns $\dot{\gamma} \leq (v + \delta)\gamma$. Hence,

(29) $\gamma(t) \leq \gamma(\tau)\, e^{(v + \delta)(t - \tau)}$ for all $t > \tau$.

Thus if $\gamma(\tau) < 0$ for some τ,

(30) $\lim_{t \to \infty} \gamma(t)\, e^{-\delta t} = -\infty$

violating (25). Similarly we can never have $q < 0$. For if $q < 0$, we must be in patterns A where $\dot{q} < (\mu + \delta)q$.

In piecing together trajectories that satisfy the necessary conditions for an optimum, it is convenient to find the stationary values of the differential equations in Table II. The trajectories can then be obtained from the backward solutions of the differential equations.

In patterns A, the only stationary point is the origin. Since $f(0) = 0$, there can be no consumption in such a steady state. Patterns C are characterized by zero consumption, so their stationary points are also unattractive as long-run equilibria. It remains to examine the stationary points of patterns B.

In order to remain in patterns B it is necessary that $q = p$, $\dot{q} = 0$. By the assumptions on f(K), there is a unique $\hat{K}(\mu) > 0$ such that $f'[\hat{K}(\mu)] = \mu$.

Then

(31) $\hat{K}'(\mu) = \frac{1}{f''(K)} < 0, \quad \hat{K}(\mu) < \tilde{K}(\mu).$

In patterns IB and IIB, $\dot{q} = 0$ when $K = \hat{K}[(\mu + \delta)p] < \hat{K}(\mu p) < \tilde{K}(\mu p)$. In pattern IIIB, $\dot{q} = 0$ when $K = \hat{K}\left[\frac{(\mu+\delta)p}{A(G)(p+\gamma)}\right] > \hat{K}[(\mu+\delta)p]$.

The stationary point in pattern IB is easily found. $\dot{G} = 0$ and $\dot{\gamma} = 0$ only when $G = \gamma = 0$. Since $\hat{K}[(\mu+\delta)p] < \tilde{K}(\mu p)$, it is possible to set $Z_C = -p\mu K > -f(K)$.

In order to remain in pattern IIB, it is also necessary that $(p + \gamma)A(G) = 1$, $\dot{\gamma}A(G) = (p + \gamma)A'(G)\dot{G} = 0$. Thus

(32) $\gamma(v + \delta)A(G) - (p + \gamma)vG\,A'(G) = 0$, or equivalently

(33) $\psi(G) \equiv (v + \delta)[1 - pA(G)] - \frac{vG\,A'(G)}{A(G)} = 0$

If $\underline{A} \le \frac{1}{p} \le \overline{A}$, there are an odd number of solutions to equation (33), if $\frac{1}{p} > \overline{A}$, there are an even number of solutions; if $\frac{1}{p} < \underline{A}$, there are no solutions. All solutions $\hat{G}(p)$ lie in the interval $0 < \hat{G}(p) \le \tilde{G}(p)$, where

$$\tilde{G}(p) \equiv \begin{cases} 0 & \text{for } 1/p \le \underline{A} \\ A^{-1}(1/p) & \text{for } \underline{A} < 1/p < \overline{A} \\ \infty & \text{for } 1/p \ge \overline{A} \end{cases}$$

Suppose we start from the initial condition

$$(34) \quad \begin{cases} K_0 = K[(\mu+\delta)p], \\ G_0 = \hat{G}(p). \end{cases}$$

Then we may set

$$(35) \quad \begin{cases} q_0 = p, \\ \gamma_0 = \frac{1}{A(G_0)} - p, \\ s = \frac{vG_0}{A(G_0)f(K_0)} \equiv \frac{(v+\delta)\gamma_0}{A'(G_0)f(K_0)(p+\gamma_0)} \\ Z_C = p[sA(G_0)f(K_0) - \mu K_0]. \end{cases}$$

Such a policy will satisfy the Pontryagin necessary conditions and we will have

$$(36) \quad \begin{cases} K(t) = K_0,\; G(t) = G_0,\; q(t) = q_0,\; \gamma(t) = \gamma_0, \text{ and} \\ V_1 = \frac{1}{\delta}\{[(1-s) + s\,pA(G_0)]f(K_0) - p\mu K_0\} \end{cases}$$

But we may also set

$$(37)\quad \begin{cases} q_0 = p \\ \gamma_0 = 0, \\ s = 0, \\ Z_C = p\mu K_0. \end{cases}$$

This policy also satisfies the Pontryagin necessary conditions and we will have

$$(38)\quad \begin{cases} K(t) = K_0, \\ G(t) = G_0\, e^{-vt}, \\ q(t) = q_0, \\ \gamma(t) = \gamma_0, \quad \text{and} \\ V_2 = \frac{1}{\delta}\left[f(K_0) - p\mu K_0\right]. \end{cases}$$

Since $A(G_0) < \frac{1}{p}$ we see that

$$(39)\quad V_2 - V_1 = \frac{s}{\delta}\left[\, 1 - pA(G_0)\, \right] f(K_0) > 0.$$

We see that if we start from a stationary point in pattern IIB, it is better to adopt a policy that leads to the stationary point in pattern IB rather than the policy which remains in IIB. It follows that there are no optimal paths that approach any stationary point in pattern IIB.

Now let us examine pattern IIIB for stationary points. In K, G space, such a point must be on the $\dot{G} = 0$ locus. To be feasible, we must have $\dot{K} = 0$ with some $Z_C \geq 0$. This condition is satisfied on segments of the $\dot{G} = 0$ locus bounded on the left by a saddle point or the origin, and on the right by a stable node. To be attainable from small initial values of K and G, the stationary point must be to the left of $(\tilde{K}, \tilde{G})$ the first intersection of $\dot{K} = 0$ with $\dot{G} = 0$ in pattern IIIC. As previously remarked, that intersection is a stable node. Its characteristic polynomial is

$$(40)\quad \begin{vmatrix} A(\tilde{G})\, f'(\tilde{K}) - \mu - \rho & A'(\tilde{G})\, f(\tilde{K}) \\ A(\tilde{G})\, f'(\tilde{K}) & A'(\tilde{G})\, f(\tilde{K}) - v - \rho \end{vmatrix}$$
$$= \rho^2 + \left[\, \mu + v - A(\tilde{G})\, f'(\tilde{K}) - A'(\tilde{G})\, f(\tilde{K})\, \right]\rho + \left[\, \mu v - \mu A'(\tilde{G})\, f(\tilde{K}) - vA(\tilde{G})\, f'(\tilde{K})\, \right] = 0.$$

Since it is a node, we must have

$$(41)\quad \mu v - \mu A'(\tilde{G})\, f(\tilde{K}) - vA(\tilde{G})\, f'(\tilde{K}) > 0,$$

or equivalently

$$(42)\quad \frac{A'(\tilde{G})\, f(\tilde{K})}{v} + \frac{A(\tilde{G})\, f'(\tilde{K})}{\mu} < 1.$$

From Table II the conditions $\dot{q} = 0$ and $\dot{\gamma} = 0$ imply

$$(43)\quad \frac{\gamma}{p} = \frac{A'(G)\, f(K)}{v + \delta - A'(G)\, f(K)} = \frac{\mu + \delta - A(G)f'(K)}{A(G)f'(K)},$$

or equivalently

(44) $$\frac{A'(G)\, f(K)}{v + \delta} + \frac{A(G)f'(K)}{\mu + \delta} = 1$$

We seek a point (K*, G*) on the $\dot{G} = 0$ locus between (0, 0) and $(\tilde{K}, \tilde{G})$.

For $K \leq \hat{K}\left[\frac{\mu + \delta}{A}\right]$ we have $\frac{A(G)f'(K)}{\mu + \delta} \geq 1$. On the other hand, at $(K, G) = (\tilde{K}, \tilde{G})$ we have by (42)

(45) $$\frac{A'(\tilde{G})f(\tilde{K})}{v + \delta} + \frac{A(\tilde{G})f'(\tilde{K})}{\mu + \delta} < \frac{A'(\tilde{G})f(\tilde{K})}{v} + \frac{A(\tilde{G})f'(\tilde{K})}{\mu} < 1.$$

Therefore, there is an odd number of stationary points (K*, G*) to the left of $(\tilde{K}, \tilde{G})$ in pattern IIIB. For simplicity of exposition, we shall assume that (K*, G*) is unique. Note that equations (26) and (44), which determine K* and G*, do not depend on the value of p. Substituting K* and G* into (43) gives the stationary value of γ as a proportional function $\gamma^*(p)$ of p.

In order to remain in pattern IIIB, it is necessary to keep q = p, $\dot{q} = 0$, $\ddot{q} = 0$. Therefore, by Table II we have

(46) $\dot{q} = (\mu + \delta)\, p - A(G)\, f'(K)(p + \gamma) = 0$, or

equivalently

(47) $$\gamma = \frac{(\mu + \delta) - A(G)\, f'(K)}{A(G)f'(K)}\, p.$$

Then substituting from Table II into

(48) $$\ddot{q} = -A'(G)f'(K)(p + \gamma)\, \dot{G} - A(G)f''(K)(p+\gamma)\dot{K} - A(G)f'(K)\dot{\gamma} = 0$$

we obtain

(49) $$\dot{K} = A(G)\, f(K) - \frac{Z_C}{p} - \mu K$$

$$= (v + \delta)\, \frac{f'(K)}{f''(K)}\left[\frac{vG\, A'(G)}{(v + \delta)A(G)} + \frac{A(G)f'(K)}{\mu + \delta} - 1\right]$$

which can be solved for Z_C. The $\dot{K} = 0$ locus defines K as a function of G,

(50) $$\check{K}(G) = \hat{K}\left\{\frac{\mu + \delta}{A(G)}\left[1 - \frac{vG\, A'(G)}{(v + \delta)A(G)}\right]\right\} > \hat{K}\left[\frac{\mu + \delta}{A(G)}\right].$$

Then

(51)
$$\check{K}(0) = \hat{K}\left(\frac{\mu + \delta}{A}\right)$$
$$\check{K}(G^*) = K^*$$
$$\check{K}(\infty) = \hat{K}\left(\frac{\mu + \delta}{\bar{A}}\right).$$

The trajectories which remain in pattern IIIB will be as shown in Figure 5.

We have established the existence of two relevant stationary points which an optimal path may reach or approach asymtotically. In pattern IB there is a stationary point at $(\hat{K}\,[(\mu + \delta)p],\ 0,\ p,\ 0)$; in pattern IIIB there is one at $(K^*, G^*, p, \gamma^*(p))$. Depending on the international price p, one of these may be dominated or eliminated. By (46) we know that

$$(52)\quad \frac{f'(K^*)}{\mu + \delta} = \frac{p}{A(G^*)[p + \gamma^*(p)]} < \frac{1}{A(G^*)} .$$

If $p < \frac{f'(K^*)}{\mu + \delta}$, then $\gamma^*(p) < \frac{1}{A(G^*)} - p$, violating the condition given in Table I for being in pattern IIIB. Thus, the second stationary point is eliminated.

If $\frac{f'(K^*)}{\mu + \delta} \leq p < \frac{1}{A(G^*)}$, starting from initial values

$$(53)\quad K_0 = K^*,\ G_0 = G^*,$$ we can set

$$(54)\quad q_0 = p,\quad \gamma_0 = \gamma^*(p).$$

Then we will have

$$(55)\quad \begin{cases} K(t) = K^*,\ G(t) = G^*,\ \tilde{q}(t) = p,\ \gamma(t) = \gamma^*(p), \text{ and} \\ V_3 = \frac{1}{\delta}\,[pA(G^*)\, f(K^*) - p\, K^*\,]. \end{cases}$$

But if we set

$$(56)\quad K(t) = K^*,\ s(t) = 0,$$

we obtain the feasible path

$$(57)\quad \begin{cases} K(t) = K^*,\ G(t) = G^*\, e^{-vt}, \\ V_4 = \frac{1}{\delta}\,[f(K^*) - p\, K^*]. \end{cases}$$

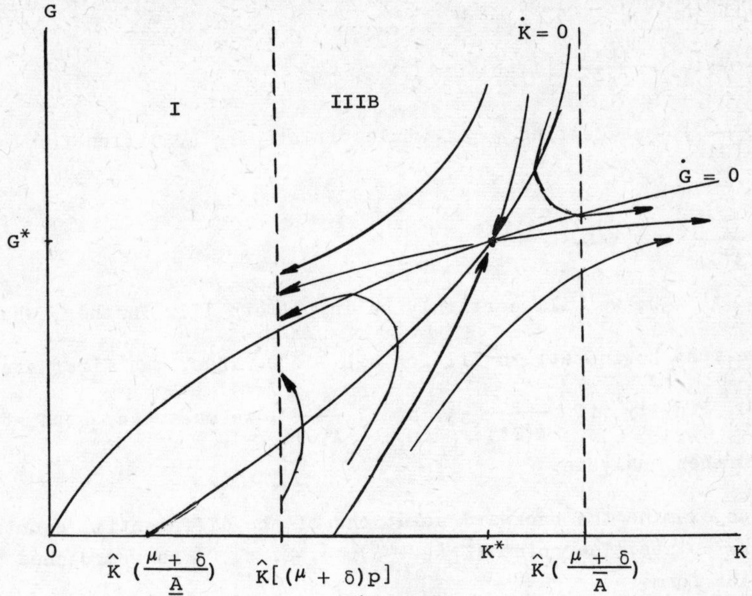

Figure 5

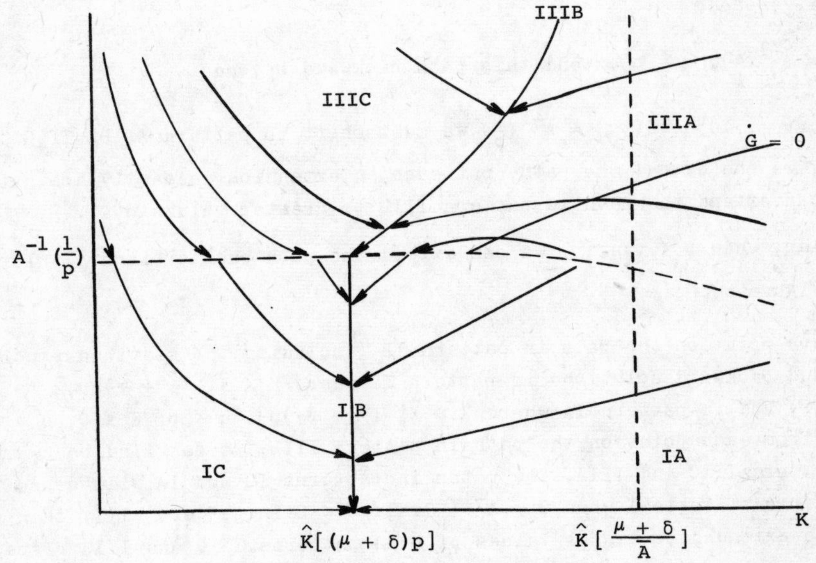

Figure 6

Although the path (57) does not satisfy the Pontryagin necessary condition, it dominates path (55) which does, since

(58) $$V_4 - V_3 = \frac{f(K^*)}{\delta} [1 - pA(G^*)] > 0.$$

If $p > \frac{1}{A(G)}$, then setting $\gamma = 0$ would violate the condition for being in pattern IB,

(59) $$\gamma < \frac{1}{A(G)} - p < 0.$$

Thus when $G > A^{-1}(\frac{1}{p})$, we will certainly be in pattern III. Further, when $p > \frac{1}{A(0)}$, we must be in pattern III for all $G \geq 0$. Thus, the first stationary point is eliminated. Finally, if $\frac{1}{A(G^*)} < p < \frac{1}{A(0)}$, we must keep both stationary points for further analysis.

Now let us examine the backward solutions of the differential equations from the two stationary points. The point $(\hat{K}[(\mu + \delta)p], 0, p, 0)$ is approached in pattern IB by paths of the form

(60) $$\begin{cases} K(t) = \hat{K}[(\mu + \delta)p], \quad G(t) = G_0 e^{-v(t - t_0)}, \\ q = p, \gamma = 0 \end{cases}$$

where $G_0 = A^{-1}(\frac{1}{p})$. To extend this path backward beyond $K(t_0) = \hat{K}[(\mu + \delta)p]$, $G(t_0) = A^{-1}(\frac{1}{p})$ we must shift to pattern IIIB. If $\dot{G} \geq 0$ at this point, we can extend the path, but such an extension is easily dominated. If $\dot{G} < 0$, we can extend the path in pattern IIIB to greater values of G. In particular, when $p < \frac{1}{A(G^*)}$, we can extend the path back arbitrarily large values of G (see Figure 6).

From each point on the path in pattern IB (including the stationary point), we can also find backward solutions in pattern IC where $K < \hat{K}[(\mu + \delta)p]$, $\dot{K} > 0$, $\dot{q} < 0$, and in pattern IA where $K > \hat{K}[(\mu + \delta)p]$, $q < p$, $\dot{K} < 0$, $\dot{q} > 0$. Similarly, from each point on the path in pattern IIIB, we can find backward solutions in patterns IIIC and IIIA. The paths in patterns IC and IA will cease to satisfy the constraints of Table I when $G = A^{-1}(\frac{1}{p})$. If $\dot{G} < 0$ in pattern III, then these paths can be extended to higher values of G in patterns IIIC and IIIA respectively.

The locus separating pattern III from pattern I will be horizontal at $G = A^{-1}(\frac{1}{p})$ to the left of the $\dot{G} = 0$ locus in pattern III. To the right of the $\dot{G} = 0$ locus, the locus separeting pattern IIIA from pattern IA will slope downward. Its exact position can only be found by comparing the values of the criterion functional (19) along two paths satisfying the necessary optimality conditions--one remaining in IA, the other starting in IIIA.

For smaller values of p, the stationary value $K = \hat{K}[(\mu + \delta)p]$ will be greater, and the region of pattern I will be larger. If $p < \frac{1}{\bar{A}}$, pattern I will cover the entire K, G plane. For larger values of p, the region of pattern I will be smaller, and there may be many initial values of K and G for which there are no solutions to the differential equations of Table II that approach the stationary point of pattern IB (see Figure 7). In fact, pattern I disappears entirely for $p > \frac{1}{\underline{A}}$.

Now let us consider trajectories which approach the terminal point $(K^*, G^*, p, \gamma^*(p))$. There will be two such paths approaching in pattern IIIB, and one each in patterns IIIA and IIIC. For initial conditions to the left of the IIIB paths, there is a path which begins with pattern IIIC and switches to pattern IIIB when (K, G) reaches a point on one of the converging IIIB paths. For initial conditions to the right of the IIIB paths, there is a path which begins with pattern IIIA and switches to pattern IIIB when (K, G) reaches a point on one of the IIIB paths (see Figure 8). These paths are invariant in K and G as p varies, although the amount of consumption at each point on the path will be proportional to p and the region in which patterns III satisfy the conditions of Table I will expand as p rises. In fact, pattern III will cover the entire K, G plane if $p > \frac{1}{\underline{A}}$, but will disappear when $p < \frac{1}{\bar{A}}$.

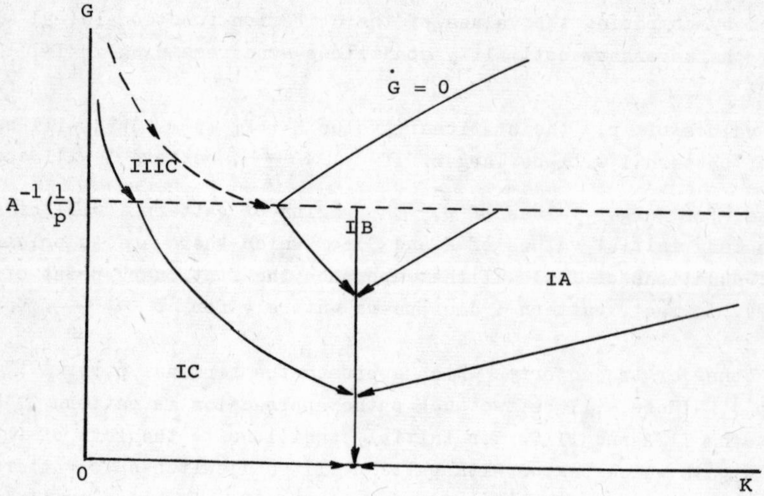

Figure 7

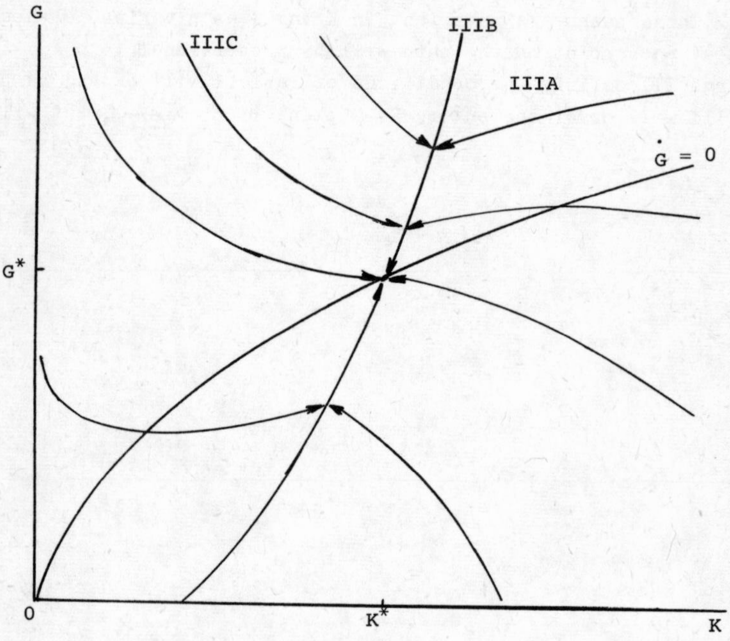

Figure 8

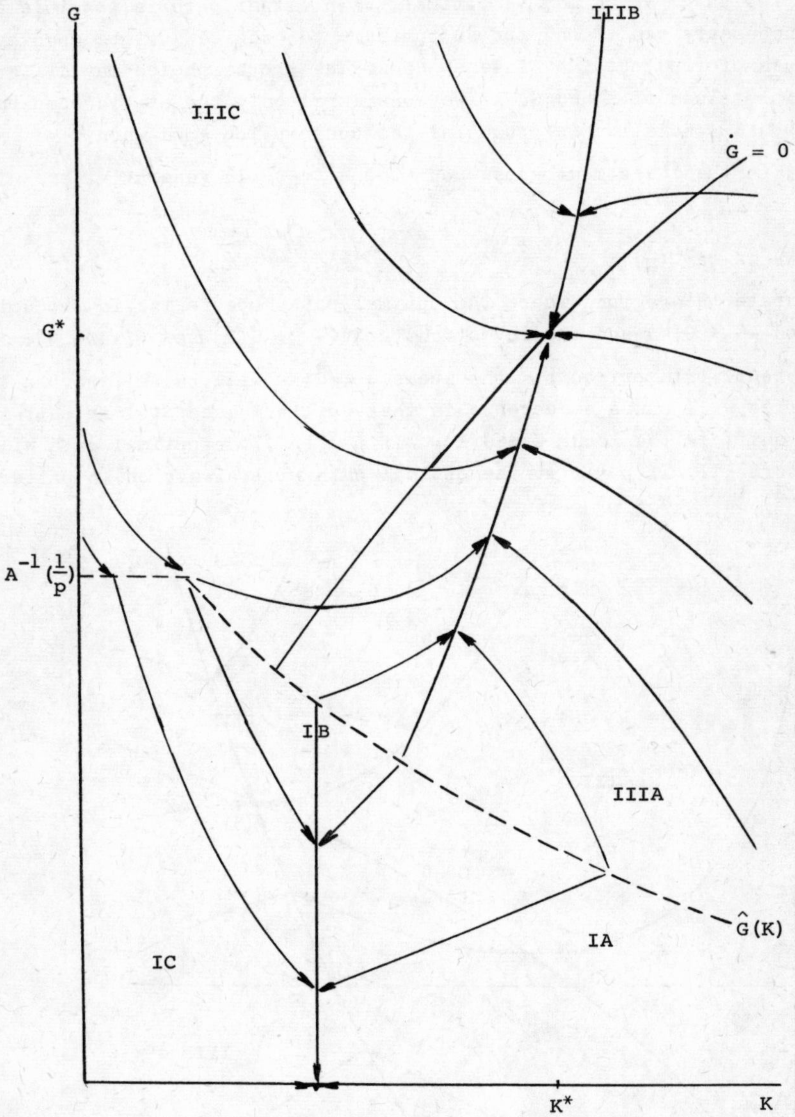

Figure 9

If

$$(61)\quad A(G^*)\left[1+\frac{\gamma^*(p)}{p}\right] > \frac{1}{p} > \underline{A},$$

then there are two paths which may be optimal. When either path is possible and satisfies all necessary conditions for an optimum, the choice must be made by comparing the values of (19) obtained in each case. The proper choice may differ for different initial values of K and G. If we considered only the static comparative advantage, we would specialize in producing the consumption good when $G < A^{-1}(\frac{1}{p})$ and in the investment good when $G > A^{-1}(\frac{1}{p})$. In general, there will be another locus,

$$(62)\quad G = \hat{G}(K) < A^{-1}(\tfrac{1}{p})$$

which will separate the regions where the optimal paths specialize in producing the consumption good--$G < \hat{G}(K)$ and the investment good--$G > \hat{G}(K)$ (see Figure 9).

Under different world prices p, the locus $G = \hat{G}(K)$ will be shifted. An increase in the world price will cause a decrease in the equilibrium capital in pattern IB and a downward shift in the locus $G = \hat{G}(K)$. If $p > 1/\underline{A}$, the optimal path will always be in pattern III. If $p < 1/\overline{A}$, the optimal path will always be in pattern I.

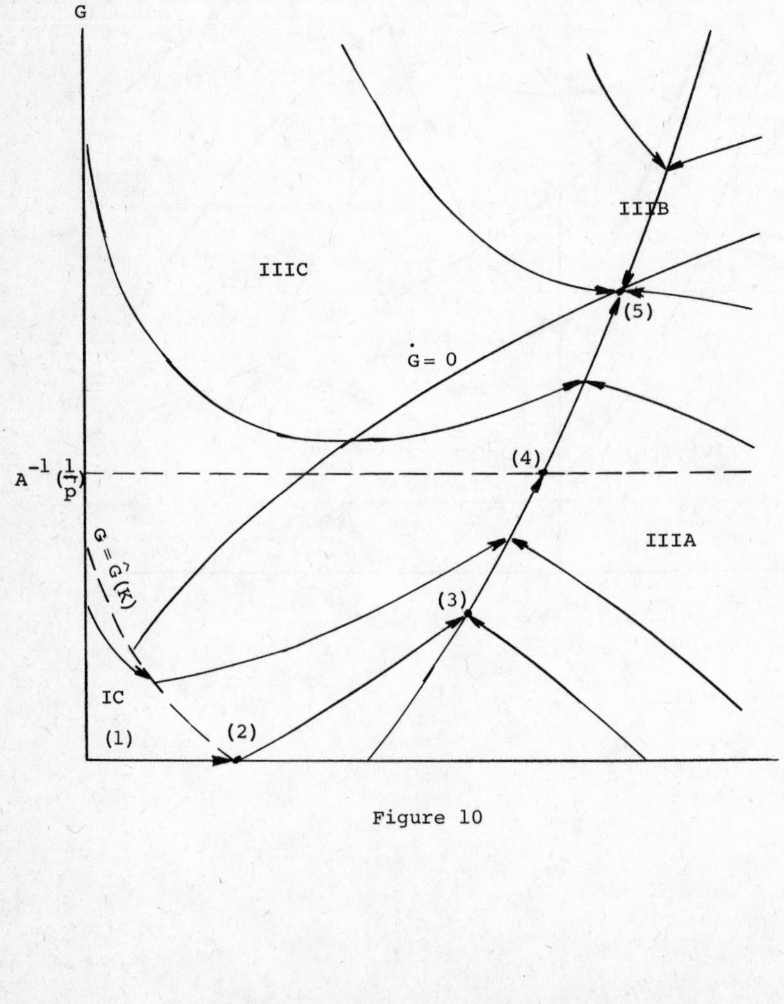

Figure 10

3. Dynamic Behavior

When the price is high but less than 1/A, we may have the situation shown in Figure 10. A country that lacks both capital and experience may go through several stages of development.

(1) At first it may concentrate on building up its capital as rapidly as possible by producing the consumption good for export.

(2) Then, despite the fact that its static comparative advantage is still in the consumption good, it may produce the investment good for domestic use.

(3) After enough capital and experience have been accumulated some of the investment good can be exported for the consumption good.

(4) Eventually the static comparative advantage shifts to the investment good.

(5) As we approach the stationary state, the comparative advantage in the investment good becomes strong enough to compensate for the static inefficiency suffered during Phases 2 and 3.

Such a path bears strong analogies to the stages of development and their corresponding trade policies as described by List[6].

When the price is somewhat lower, we will have the situation shown in Figure 9. Here a country with an initial experience greater than $A^{-1}(\frac{1}{p})$ should follow its comparative advantage by specializing in the investment good. A country with initial experience less than $\hat{G}(K)$ should follow its comparative advantage by specializing in the consumption good. A country with an intermediate initial level of experience

$\hat{G}(K) < G_0 < A^{-1}(\frac{1}{p})$ should specialize in the investment good despite a temporary comparative disadvantage. As it builds up experience, it will move into the first category.

---

6. *Op. cit.*, p. 108. "Die Geschichte lehrt uns endlich, wie Nationen, die mit allen zur Erstrebung des hoechsten Grades von Reichtum und Macht erforderlichen Mitteln von der Natur ausgestattet sind, ohne mit ihrem Bestreben in Widerspruch zu geraten, nach Maßgabe ihrer Fortschritte mit ihren Systemen wechseln koennen und muessen, indem sie durch freien Handel mit weiter vorgerueckten Nationen sich aus der Barbarei erheben und ihren Ackerbau emporbringen, hierauf durch Beschraenkung das Aufkommen ihrer Manufakturen, ihrer Fischereien, ihrer Schiffahrt und ihres auswaertigen Handels befoerdern und endlich, auf der hoechsten Stufe des Reichtums und der Macht angelangt, durch allmaehliche Rueckkehr zum Prinzip des freien Handels und der freien Konkurrenz auf den eigenen wie auf den fremden Maerkten, ihre Landwirte, Manufakturisten und Kaufleute gegen Indolenz bewahren und sie anspornen, das erlangte Uebergewicht zu behaupten. Auf der ersten Stufe sehen wir Spanien, Portugal und Neapel stehen, auf der zweiten Deutschland und Nordamerika; den Grenzen der letzten Stufe scheint uns Frankreich nahe zu sein; erreicht hat sie zur Zeit allein Großbritanien.

4. Development in a Multi-Country World

It may be instructive to examine the development process in a world made up of a large number n of countries like the one whose optimal behavior we have analyzed. Let us assume that each country adopts a trade and growth policy that would be optimal on the assumption that the world price remain constant at its current level. The initial values of the ith country's capital and experience will be K_0^i, G_0^i (i = 1,2, ..., n). We assume that K_0^i, G_0^i are small for all countries. As long as all countries are to the left of the IIIB locus in Figure 10, the price must be high enough that pattern IC disappears. Each country will then choose to be self-sufficient, producing and using only the investment good.

Eventually, the most advanced country will reach pattern IIIB and begin to offer part of its output of the investment good for export. The price must then begin to fall. As it falls, the quantity of the investment good exported by the most advanced country is fixed, hence the effective demand for imports of the consumption good falls proportionally to the price. At the same time pattern IC appears near the origin and expands until it includes one or more of the least advanced countries. With falling price, patterns I expand at the expense of patterns III and pattern IC expands at the expense of pattern IA. Thus, the supply of exports of the consumption good rises in a step function as the price falls. There will be a unique short-run equilibrium price p*, which may leave a marginal country teetering back and forth in indecision between patterns I and patterns III.(See Figure 11.)

With the passage of time the more advanced countries in pattern IIIB will increase their exports of the investment good. Exports of the investment good will also receive a boost every time a new country moves from pattern IIIC to IIIB. Countries in pattern IC will increase exports of the consumption good until they reach pattern IB, where there will be a sharp drop. There is considerable leeway for minor price fluctuations, but the general tendency of the price will be downward.

In the later stages of the process, most of the advanced countries will have reached pattern IIIB and most of the less advanced countries will have reached pattern IB. The margin will be thinned out, since countries on both sides will be moving away. The limits within which the price may fluctuate without causing any country to drastically alter its plan will be widened. Ultimately, the system approaches an equilibrium in which all the advanced countries achieve $(K^i, G^i) = (K^*, G^*)$, and all the other countries achieve $(K^i, G^i) = (\hat{K}[(\mu + \delta)p], 0)$. The equilibrium value of p* will depend on how many countries wind up in each category: the greater the number of advanced countries, the lower the price. Thus, the greater the number of advanced countries, the less advantageous it is to be an advanced country. The number of advanced countries is confined to certain broad limits by the requirement that the advantage of being an advanced country be greater than the temporary benefit of moving from an advanced to an unadvanced state and less than the temporary cost of moving from an unadvanced to an advanced state. Within these limits, the precise equilibrium approached will depend on the historical accidents that determine the distribution of initial values (K_0^i, G_0^i).

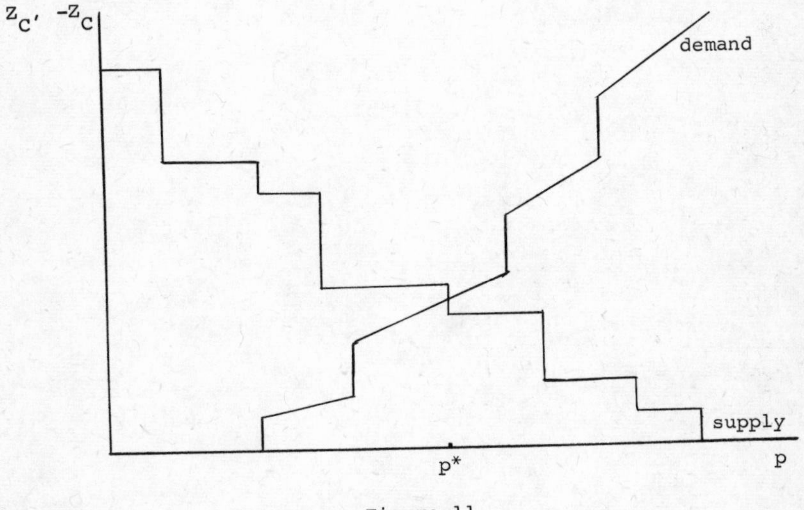

Figure 11

Multi-level Approach to the Large-scale Control Problem

A. Straszak

I. Statement of the Problem

Let us consider a multivariable control problem. From the theory of the optimal control [1], [5] we know that it is possible to find for some speciale case the so-called "feed-back" solution of the optimal control problem

$$u = c\,(x)$$

where

u - is a n-dimensional optimal control vector which minimized the index of performance

$$I = \int_{t_o}^{t_1} g(x,u)\,dt$$

x - is a n-dimensional state vector of the controlled process

$$\dot{x} = f\,(x,u)$$

Let us introduce the complexity measure of the optimal controller.
Realization of the multivariable controller needs some list of the data processing operation.
For example, if the optimal controller is linear

$$u = C\,x$$

where C - n x n - matrix with constants nonzero parameters
than list of the data processing operation is following :

1/ n x n multiplication by constant c_{ij}
2/ n summations

if the optimal controller is nonlinear, than the list of the data processing operation may include multiplication variables by variables, generation of the nonlinear function and so on. In general, for the given control law it is necessary to have the list of the data processing operation which include

1. m_α operation α
2. m_β operation β
3. m_γ operation γ

\- - - - - - - - - - - -

and so on.

Let us define the complexity of the controller as the complexity of the list of the data processing operation which is necessary for given control law.

Let us introduce the complexity function Z

$$Z = m_\alpha k_\alpha + m_\beta k_\beta + m_\gamma k_\gamma + \dots$$

where

k_α - complexity of the operation α

k_β - complexity of the operation β

k_γ - complexity of the operation γ

For linear controller

$$Z = n^2 k_c + n k_\Sigma$$

where

k_c - complexity of the multiplication by constant

k_Σ - complexity of the summation.

In general for multivariable controller [9], [10]

$$Z = \sum_{i=a}^{r} n^{p_i} k_i$$

$$p_i = 1,2,3,\dots$$

For large-scale control problems

$$Z_c > Z_o$$

where Z_c - complexity of the control law

Z_o - available complexity of the list of data processing operation.

Therefore the following control problem arises:

Given the complexity of the list of the data processing $Z_o < Z_c$, find the new control law c_m and new list of the processing such that $Z_{c_m} \leq Z_o$ and $I = I\,|x, c\,(x)| - I\,|x, c_m\,(x)|$ will be as small as possible.

II. Some properties of function Z

Let us suppose that

$$n = \sum_{j=1}^{m} n_j \qquad m = 2,3,\dots,1 < n$$

there exists n such that

$$Z = \sum_{i=1}^{r} n^{p_i} k_i = \sum_{i=1}^{r} \left(\sum_{j=1}^{m} n_j \right)^{p_i} k_i \geq \sum_{j=1}^{m} \sum_{i=1}^{r} n_j^{\;p_i} k_i =$$

$$= \sum_{j=1}^{m} Z_j \;.$$

It means that by decomposition of the control problem it is possible to reduce the complexity of the control law.

It also can be shown, that exist such n, m, $u_o \geq 1$, $u_j \geq 1$ and $\lambda \geq 1$ that

$$Z = \sum_{i=1}^{r} n^{p_i} k_i \quad \sum_{j=1}^{m} u_j \sum_{i=1}^{r} n_j^{\;p_i} k_i + u_o \sum_{i=j}^{r} \lambda m^{\,p_i} k_i$$

It means that by the decomposition of the control problem it is possible to reduce the complexity of the control law even if subproblems are more complicated and even if it is necessary to introduce new controller for coordination between subcontrollers [7] [10] .

III. Supervisory control

Using the decomposition of the control problem due to the complexity of the control law, in general, we loose the global optimality of the control system, therefore it is necessary to improve the quality of the control by introducing a new supervisory level of the control which will compensate the influence between subsystem or will allocate the resource between the subsystems.

Let us suppose that $I_i = \int_{t_o}^{t_1} g_i(x_i, u_i)\, dt$ is the performance index of the subsystem "i", than to simplify the supervisory control law we may use a new global performance index in following form

$$I_g = \sum_{j=1}^{m} \min_{i \in R} I_i \; (x_{io}, u_i)$$

where R - resources or compensation factor which are allocated by the supervisory controller to the optimal controller of the subsystem "i".

Therefore the goal for the supervisory controller can be following :

Minimize $$I_g = \sum_{j=1}^{m} \min \; I_i \; [\, x_{io}, R(u_i)]$$

subject to the constraints

$$R_1 (u_1) + R_2 (u_2) + \ldots + R_m(u_m) \leq R_o$$

and

$$R_i (u_i) \geq 0$$

which can be solved by using mathematical programming machinery and particularly by using also decomposition and coordination procedures [2] [3] [6] . Of course by such procedure it is possible, in general, to obtain only the suboptimal solution.
It was shown [7] , [8] , [10] , that such approach can be used to the minimum time and minimum energy control problems.

IV. Multi-level control structure

Using the results of the foregoing sections it is possible now to solve the original problem.
Let us estimate the function Z for our control problem.

If $Z_c > Z_o$, than our large-scale control problem must be decomposed into subproblems and probably it will be necessary to use the supervisory controller or even decomposition and coordination of the supervisory control such that it will be necessary to use a several level of control, therefore it will be necessary to estimate the function Z_s for supervisory control and higher level of control.

Let $Z_w, Z_{s2}, Z_{s3}, \ldots, Z_{sk}$ functions Z for first, second,..., k-level of control than admissible control structures can be obtained from the following inequality

$$\sum_{c=1}^{m_1} Z_{wi} + \sum_{a=1}^{m_2} Z_{s_{2_a}} + \sum_{b=1}^{m_3} Z_{s_{2_b}} + \ldots + \sum_{h=1}^{m_n} Z_{s_{k_h}} \leq Z_o$$

From admissible multi-level control structures we choose control structures for which $k, m_h, m_{h-1}, \ldots m_2, m_1$ are minimum. Than for this control structure we define the control law for first, second and k - level controllers [7] [10] .

References

1. Chang A. : An optimal regulator problem. J. SIAM Ser. A. Control 1964

2. Dantzig G.B. : Linear programming and extension. Princeton University Press. Princeton 1963.

3. Karlin S. : Mathematical Methods and Theory in Games, Programming and Economics. Pergamon Press. London 1959

4. Mesarovič M.D. and others : Adavances in multi-level control. IFAC Tokyo Symposium on System Engineering. Tokyo, 1965.

5. Pontriagin L.S. and others : Mathematical Theory of the Optimal Process. Fizmatgiz, Moscow, 1961.

6. Pearson J.D., Reich S. : The Decomposition of Large Optimal Control Problems. Case Institute of Technology. Report NO 98-A-66-90, July, 1966.

7. Straszak A. : On the Structure Synthesis Problem in Multi-level Control Systems. Proc. IFAC Tokyo Symposium on System Engineering, Tokyo, 1965.

8. Straszak A. : Suboptimal Supervisory Control. Functional Analysis and Optimization. Academc Press. New York, 1966.

9. Straszak A. : Optimal and Sub-optimal Multivariable Control Systems with Controller Cost Constraint. Proc. III IFAC Congress, London, 1966

10. Straszak A. : Control Methods for Large-scale Multivariable Dynamic Processes. Raport No 60, Inst.for Autom.Control, Warsaw, 1967, /in Polish/ .

Offsetdruck: Julius Beltz, Weinheim/Bergstr.

Lecture Notes in Operations Research and Mathematical Economics

Vol. 1: H. Bühlmann, H. Loeffel, E. Nievergelt, Einführung in die Theorie und Praxis der Entscheidung bei Unsicherheit. IV, 125 Seiten 4°. 1967. DM 12,– / US $ 3.00

Vol. 2: U. N. Bhat, A Study of the Queueing Systems M/G/1 and GI/M/1. VIII, 78 pages. 4°. 1968. DM 8,80 / US $ 2.20

Vol. 3: A. Strauss, An Introduction to Optimal Control Theory. VI, 153 pages. 4°. 1968. DM 14,– / US $ 3.50

Vol. 4: Einführung in die Methode Branch and Bound. Herausgegeben von F. Weinberg. VIII, 159 Seiten. 4°. 1968. DM 14,– / US $ 3.50

Vol. 5: L. Hyvärinen, Information Theory for Systems Engineers. VIII, 295 pages. 4°. 1968. DM 15,20 / US $ 3.80

Vol. 6: H. P. Künzi, O. Müller, E. Nievergelt, Einführungskursus in die dynamische Programmierung. IV, 103 Seiten. 4°. 1968. DM 9,– / US $ 2.25

Vol. 7: W. Popp, Einführung in die Theorie der Lagerhaltung. VI, 173 Seiten. 4°. 1968. DM 14,80 / US $ 3.70

Vol. 8: J. Teghem, J. Loris-Teghem, J. P. Lambotte, Modèles d'Attente M/G/1 et GI/M/1 à Arrivées et Services en Groupes. IV, 53 pages. 4°. 1969. DM 6,– / US $ 1.50

Vol. 9: E. Schultze, Einführung in die mathematischen Grundlagen der Informationstheorie. VI, 116 Seiten. 4°. 1969. DM 10,– / US $ 2.50

Vol. 10: D. Hochstädter, Stochastische Lagerhaltungsmodelle. VI, 269 Seiten. 4°. 1969. DM 18,– / US $ 4.50